Lecture Notes in Control and Information Sciences 225

Editor: M. Thoma

Springer-Verlag London Ltd.

A.S. Poznyak and K. Najim

Learning Automata and Stochastic Optimization

Springer

Authors

A.S. Poznyak
CINVESTAV, Instituto Polytecnico Nacional México,
Departamento do Ingenieria Electrica, Seccion de Control Automatico
A.P. 14-740, 07000 México D.F., México

K. Najim
Institut National Polytechnique de Toulouse,
Toulouse, France

ISBN 978-3-540-76154-9

British Library Cataloguing in Publication Data
Pozniak, A. S. (Aleksandr Semenovich)
Learning automata and stochastic optimization. - (Lecture notes in control and information sciences ; 225)
1.Mathematical optimization 2.Machine theory 3.Stochastic processes
I.Title II.Najim, Kaddour
519.2'3
ISBN 978-3-540-76154-9 ISBN 978-3-540-40938-0 (eBook)
DOI 10.1007/978-3-540-40938-0

Library of Congress Cataloging-in-Publication Data
A catalog record for this book is available from the Library of Congress

Typesetting: Camera ready by authors

69/3830-543210 Printed on acid-free paper

To Tatyana and Michelle

Contents

0.1 Notations

The upperscript convention will be used to index the discrete-time step. Throughout this book we use the following notation:

$\arg\min_x f(x)$	minimizing argument of $f(x)$
B_{n-1}	$\sigma(u_1, ..., u_n; \omega_1, ..., \omega_{n-1})$
$c_n(i)$	conditional mathematical expectations of the environment responses
$\widetilde{c}(i)$	arithmetic averages limit of $c_n(i)$
d_i	diameter of the subset X_i
E	expectation operator
$en\ X$	envelope of the set X
$e^{N(j)}$	$= (1, ..., 1) \in R^{N(j)}$
$e^{N(j)}(u_n)$	$= (0, 0, ..., 0, 1, 0, ..., 0)^T \in R^{N(j)}$ (the sth component of this vector is equal to 1 if $u_n = u(s)$ and the other components are equal to zero)
$F_x(x)$	distribution function of the vector x
$\mathcal{F}_n$	σ-algebra (σ-field:a set of subsets of Ω for which probabilities are defined))
$f_0(.)$	objective function
$f_j(.)$	constraints ($j = 1, ..., m$)
L_i	Lipschitz constants
$L(x, \lambda)$	Lagrange function
$L_{\mu,\delta}(p, \widetilde{u})$	penalty function
N	number of automaton actions
$N(j)$	number of actions of the subset $V(j)$
P	probability measure
p_n	probability distribution at time n
$p_n(\alpha)$	probability associated with the optimal action $u(\alpha)$
p^*	optimal strategy
q_n	probability distribution defined over all possible actions subsets
$\mathcal{R}$	$= (\Omega, \mathcal{F}, P)$ a probability space
U	set of actions
u_n	automaton output (action) at time n
$u(\alpha)$	optimal action
$\widetilde{u}_j$	slack variables ($j = 1, ..., m$)
V	set of actions

$V(j)$	j^{th} subset of actions
$V_0(.)$	objective function
$V_j(.)$	constraints ($j = 1, ..., m$)
W_n	Lyapunov function
$\{X_i\}$	quantification of the compact set X
x^*	optimal solution
y_n	observation at time n
γ_n	correction factor
$\delta(.)$	$\delta-$ function
ζ_n^j	multi-teacher environment responses at time n ($j = 1, ..., m$)
$\lambda(j)$	Lagrange multipliers ($j = 1, ..., m$)
μ	penalty coefficient
ξ	environment response
$\widetilde{\xi}$	normalized environment response
$O(x)$	$O(x)/x \rightarrow 0$ bounded when $x \rightarrow 0$
$o(x)$.	$o(x)/x \rightarrow 0$ when $x \rightarrow 0$
Π	projection operator
$\Phi_n^0(.)$	objective function
$\Phi_n^j(.)$	constraints ($j = 1, ..., m$)
$\sigma_n^2(i)$	conditional variances of the environment responses
Φ_n	loss function at time n
Ω	basic space (events space)
ω_n	observation noise at time n

0.2 Introduction

In the last decades there has been a steadily growing need and interest in computational methods for solving optimization problems with or without constraints. They play an important role in many fields (chemistry, mechanic, electrical, economic, etc.). Optimization techniques have been gaining greater acceptance in many industrial applications. This fact was motivated by the increased interest for improved economy in and better utilization of the existing material resources. Euler says: "Nothing happens in the universe that does not have a sense of either certain maximum or minimum". In this book we are primarily concerned with the use of learning automata as a tool for solving many optimization problems. Learning systems have made a significant impact on many areas of engineering problems including modelling, control, optimization, pattern recognition, signal processing and diagnosis. They are attractive and provide interesting methods for solving complex nonlinear problems characterized by a high level of uncertainty. Learning systems are expected to provide the capability to adjust the probability distribution on-line, based on the environment response. They are essentially feedback systems. The optimization problems are modeled as that of learning automaton or a hierarchical structure of learning automata operating in a random environment. We report new and efficient techniques to deal with different kinds (unconstrained, constrained) of stochastic optimization problems. The main advantage of learning automata over other optimization techniques is its general applicability, i.e., there are almost no condition concerning the function to be optimized (continuity, differentiability, convexity, unimodality, etc.).

This book presents a solid background for the design of optimization procedures. Ample references are given at the end of each chapter to allow the reader to pursue his interest further. Several applications are described throughout the book to bridge the gap between theory and practice. Stochastic processes as martingales have extensive applications in stochastic problems. They arise naturally whenever one needs to consider mathematical expectations with respect to increasing information patterns. They will be used as well as Lyapunov functions to state several theoretical results concerning the convergence and the convergence rate of learning systems. The Lyapunov approach offers a shortcut to proving global stability (convergence) of dynamical systems. A Lyapunov function often measures the energy of a physical system.

This book consists of five Chapters. Chapter one deals with stochastic optimization problems. It is shown that stochastic optimization problems (with or without constraints) can be reduced to optimization problems on finite sets and, then solved using learning automata (operating in S-model environments, i.e., continuous environment response) which are essentially sequential machines. An introduction to learning automata and several notions and definitions are presented in Chapter two. Both single and hierarchical structures of learning automata are considered. It is shown that the synthesis of several reinforcement schemes can be associated with the optimization of some functional. Chapters one and two constitute the core of this book. Chapter three is concerned with stochastic unconstrained optimization problems. Learning automata with fixed and changing number of actions are used as a tool for solving this kind of optimization problems. A transformation called "normalization procedure" is introduced to ensure that the environment response (automaton input) belongs to the unit segment.

Several analytical results dealing with the convergence characteristics of the optimization algorithms are stated. The stochastic constrained optimization problems are treated in Chapter four. This stochastic optimization problem is formulated and solved as the behaviour of variable-structure stochastic automata operating in multi-teacher environments. Two approaches are considered: Lagrange multipliers and penalty functions. The properties of the optimal solutions are presented. The convergence of the derived optimization algorithms is stated and the associated convergence rates are estimated. Chapter five deals with the optimization of nonstationary (time-varying, etc.) functions which can encountered in several engineering problems. These optimization problems are associated with the behaviour of learning automata with continuous inputs (S-model environment) operating in nonstationary environments.

Numerical examples illustrate the performance of the optimization algorithms described in chapters three, four and five.

Two appendices end this book. The first one contains several lemmas and their proofs. the second one contains same definitions and properties concerning different kinds of convergence and martingales. A detailed table of contents provide a general idea of the scope of the book.

The content of this book, in the opinion of the authors, represent an important step in the systematic use of learning automata concepts in stochastic optimization problems.

We would like to thank our doctoral students: E. Ikonen and E. G. Ramires who have carried out most of the simulation studies presented in this book.

Professor A.S. Poznyak
Instituto Polytecnico Nacional
México, Mexico

and

Professor K. Najim
Institut National Polytechnique
de Toulouse, France

1

Stochastic Optimization

1.1 Introduction

Because of its practical significance, there has been considerable interest in the literature on the field of optimization. This field has seen a tremendous growth in several industries (chemistry, data communication, etc.). Any problem of optimization concerns the minimization of some function over some set which can be specified by a number of conditions (equality and inequality constraints) [6]-[7]. Maximization can be converted to minimization by a change of sign.

The optimization problems associated with real systems involves a high level of uncertainty, disturbances, and noisy measurements. In several optimization problems there exists no information (probability distribution, etc.) or if there exists it is incomplete.

Learning systems are adaptive machines [8]-[5]. They improve their behaviour with time and are useful whenever complete knowledge about an environment is unknown, expensive to obtain or impossible to quantify.

A learning automaton is connected in a feedback loop to the random medium (environment), where the input to one is the output to the other. At every sampling period (iteration), the automaton chooses an action from a finite action set, on the basis of the probability distribution. The selected action causes the reaction of the environment, which in turn is the input signal for the automaton (i.e., the environment establishes the relation between the actions of the automaton and the signal received at its input which can be binary or continuous). With a given reinforcement scheme, the learning automaton recursively updates its probability distribution on the basis of the environment response.

One of our first goals in this chapter will be to present the various stochastic optimization problems in an unified manner and to show how learning automata can be used to solve efficiently this kind of optimization problems.

Optimization methods based on learning systems belong to the class of random search techniques. The main advantage of random search over other direct search techniques is its general applicability, i.e., there are almost no conditions concerning the function to be optimized (continuity, unimodality, convexity, differentiability, etc.).

The standard programming problem will be stated in the next section.

1.2 Standard deterministic and stochastic nonlinear programming problem

In this section, our main emphasis will be on the standard programming problem [5]-[6] which can be stated as follows:

$$\inf_{x} f_0(x) \tag{1.1}$$

$$f_j(x) \leq 0, \ (j = 1, ..., m) \tag{1.2}$$

$$x \in X \subseteq R^M \tag{1.3}$$

The function $f_0(x)$ is an objective function, sometimes also called a criterion function or cost function. In the case of unconstrained problem the optimal solution requires the gradient at the extremum of $f_0(x)$ to be null.

Notice that equalities constraints of the form

$$\phi_s(x) = 0, \ (s = 1, ..., m_e)$$

can be transformed into inequalities constraints

$$f_j(x) \leq 0, \ (j = m+1, ..., m+2m_e)$$

where

$$f_j(x) := \phi_j(x), \ (j = m+1, ..., m+m_e)$$
$$f_j(x) := -\phi_j(x), \ (j = m+1, ..., m+m_e)$$

So, hereafter we will deal with the standard programming problem (1.1)-(1.3).

The set X may be open or close and can have different nature which are usually connected with the physical problem formulated by (1.1)-(1.3).

The standard programming problem is said to be discrete programming problem if the set X is discrete (finite set). It is said to be continuous programming problem if the set X is continuous (compact). The function $f_j(x)$ $(j = 0, ..., m)$ involved in (1.1) and (1.2) are assumed to be continuous, Lipschitzian and twice differentiable.

In deterministic nonlinear programming problems, it usually assumed that x_n and y_n are available at time n, where y_n is defined as follows

- zero-order methods

$$y_n^T := (f_0(x_n), f_1(x_n), ..., f_m(x_n)) \in R^{m+1}$$

- first-order methods

$$y_n^T := (\nabla^T f_0(x_n), \nabla^T f_1(x_n), ..., \nabla^T f_m(x_n)) \in R^{(m+1)M}$$

Our main objective is to construct an optimization procedure (strategy), i.e.,

$$x_n = x_n(y_1, ..., y_n) \in X \tag{1.4}$$

to solve the following constrained optimization problem

$$f_0(x_n) \underset{n\to\infty}{\to} \inf_{x\in X} f_0(x) \tag{1.5}$$

subject to

$$\overline{\lim_{n\to\infty}} f_j(x_n) \le 0, \ (j = 1, ..., m)$$

It is usually assumed that all the functions $f_j(x_n)$ $(j = 1, ..., m)$ are convex and enough smooth.

Among the methods of the first-order, the Arrow-Hurwicz-Uzawa method [5] is commonly used. It is the recurrent version of the gradient technique applied to the corresponding Lagrange function using the Slater's condition [7]-[8]:

$$\exists\, \overline{x} \in X : f_j(\overline{x}) < 0, \ (j = 1, ..., m) \tag{1.6}$$

In practice the function to be optimized $(f_0(x))$ as well as the constraints $(fj(x))$ display some random characteristics (incomplete knowledge). In this case the optimization problem have to be formulated as a stochastic programming problem.

1.3 Standard stochastic programming problem

A key feature of many practical problems is the presence of disturbances on both optimized and constraints functions. The stochastic programming problem is defined in terms of the minimization of

$$\inf_x \mathbf{E}\{f_0(x,\omega)\} \tag{1.7}$$

subject to

$$\mathbf{E}\{f_j(x,\omega)\} \le 0, \ (j = 1, ..., m) \tag{1.8}$$

$$x \in X \subseteq R^M \tag{1.9}$$

where $f_j(x,\omega)$ $(j = 1, ..., m)$, for each fixed $x \in X$, are Borel functions (random variables) defined on a probability space $(\Omega, \mathcal{F}, P)$ and $\omega \in \Omega$.

The probability space $(\Omega, \mathcal{F}, P)$ is defined as follows:

Ω is the space of elementary events

$\mathcal{F}$ is the σ-algebra constructed from all subsets of Ω

P is a probability measure defined on the measurable space $(\Omega, \mathcal{F})$.

Similarly to the standard deterministic programming problem, the observation y_n is defined as follows:

- zero-order methods

$$y_n^T(\omega) := (f_0(x_n,\omega), f_1(x_n,\omega), ..., f_m(x_n,\omega)) \in R^{m+1} \tag{1.10}$$

- first-order methods

$$y_n^T(\omega) := \left(y_{n,0}^T(\omega), y_{n,01}^T(\omega), ..., y_{n,m}^T(\omega)\right) \in R^{(m+1)M} \tag{1.11}$$

where

$$\mathbf{E}\left\{y_{n,j}^T(\omega) \mid \mathcal{F}_{n-1}\right\} \overset{a.s.}{=} \nabla\phi_j(x_n),\ (j = 0, ..., m) \tag{1.12}$$

$$\mathcal{F}_{n-1} := \sigma - \text{algebra generated by } (y_1(\omega), ..., y_{n-1}(\omega))$$

$$\phi_j(x_n) := \mathbf{E}\{f_j(x,\omega)\}\ (j = 0, .., m) \tag{1.13}$$

The standard assumptions concerning the observation vector $y_n(\omega)$ are the following

$$\mathbf{E}\left\{\|y_n(\omega)\|^2 \mid \mathcal{F}_{n-1}\right\} \overset{a.s.}{\leq} C_0 + C_1 \|x_n\|^2$$

In view of (1.12), the observation vector can be interpreted as the gradient of $\{\nabla\phi_j(x_n), j = 0, ..., m\}$ disturbed by zero conditional mathematical expectation noise (vector) $\xi_n^j = \xi_n^j(\omega)$ $(j = 0, ..., m)$, i.e.,

$$y_n^T(\omega) = \left[\nabla^T\phi_0(x_n) + (\xi_n^0(\omega))^T; ...; \nabla^T\phi_m(x_n) + (\xi_n^m(\omega))^T\right] \tag{1.14}$$

where

$$\mathbf{E}\left\{\xi_n^j(\omega) \mid \mathcal{F}_{n-1}\right\} \overset{a.s.}{=} 0,\ (j = 0, .., m) \tag{1.15}$$

If the statistics characteristics of all the random functions $f_j(x,\omega)$ $(j = 0, .., m)$ are assumed to be known for all $x \in X$, then the standard stochastic optimization problem (1.7)-(1.9) can be rewritten as follows:

$$\phi_0(x) = \int_{\Omega} f_0(x,\omega)F(d\omega) \to \inf_{x\in X} \tag{1.16}$$

$$\phi_j(x) = \int_{\Omega} f_j(x,\omega)F(d\omega) \leq 0\ (j = 0, .., m) \tag{1.17}$$

where the distribution function is assumed to be known and the integrals in (1.16) and (1.17) are defined in the Lebesgue-Stieltjes sense.

Remark 1 *The "chance constraints" [9]-[10]*

$$\mathbf{P}\left\{\omega : \phi_j(\zeta_n^0, \zeta_n^1, ..., \zeta_n^m) \geq \alpha_j\right\} \leq \beta_j \tag{1.18}$$

belong to the class of constraints (1.3). Indeed, this chance constraint can be written as follows:

$$\mathbf{E}\left\{\zeta_n^j\right\} \leq 0 \tag{1.19}$$

where

$$\zeta_n^j = \chi\left\{\phi_j(\zeta_n^0, \zeta_n^1, ..., \zeta_n^m) \geq \alpha_j\right\} - \beta_j \tag{1.20}$$

and the indicator function $\chi(\cdot)$ *is defined as follows*

$$\chi(\mathcal{A}) = \begin{cases} 1 \text{ if } \mathcal{A} \text{ is true} \\ 0 \text{ otherwise} \end{cases}$$

Remark 2 *Substituting (1.20) into (1.3) leads to*

$$\limsup_{n\to\infty} \frac{1}{n}\sum_{t=1}^{n} \chi\left\{\phi_j(\zeta_t^0, \zeta_t^1, ..., \zeta_t^m) \geq \alpha_j\right\} \leq \beta_j \tag{1.21}$$

In the stationary case, i.e., the probability distribution of the vector $(\zeta_t^0, \zeta_t^1, ..., \zeta_t^m)^T$ *is stationary, and in view of the strong law of large numbers [11]-[12], it follows that inequalities (1.21) and (1.18) are equivalent, i.e.,*

$$\limsup_{n\to\infty} \frac{1}{n}\sum_{t=1}^{n} \chi\left\{\phi_j(\zeta_t^0, \zeta_t^1, ..., \zeta_t^m) \geq \alpha_j\right\} =$$

$$= \limsup_{n\to\infty} \frac{1}{n}\sum_{t=1}^{n} \mathbf{E}\left\{\chi\left\{\phi_j(\zeta_t^0, \zeta_t^1, ..., \zeta_t^m) \geq \alpha_j\right\}\right\} =$$

$$= \mathbf{P}\left\{\omega : \phi_j(\zeta_n^0, \zeta_n^1, ..., \zeta_n^m) \geq \alpha_j\right\} \leq \beta_j$$

In such statement, this problem is equivalent to the standard deterministic programming problem (1.1)-(1.3). Before to apply the deterministic optimization techniques, it necessary to calculate the integrals in (1.16) and (1.17). These integrals involve complex operations and can not be exactly calculated. Several approaches have been proposed to avoid this problem. One approach consists of using the average of the observations $\{y_n(\omega)\}$ given by (1.10), i.e.,

$$\overline{\lim_{n\to\infty}} \frac{1}{n}\sum_{t=1}^{n} y_n^0(\omega) \overset{a.s.}{\to} \inf_{\{x_n\}} \tag{1.22}$$

$$\overline{\lim_{n\to\infty}} \frac{1}{n}\sum_{t=1}^{n} y_n^j(\omega) \overset{a.s.}{\leq} 0, \ (j = 0, .., m) \tag{1.23}$$

where $\{x_n\}$ belongs to the class of all realizable strategies (1.4).

This stochastic programming problem (1.22)-(1.23) leads to highly computational efforts. In fact, it is necessary to solve it (parallel optimization) for different realizations.

Another approach is based on the implementation of stochastic approximation techniques [13]-[14]-[15] using the observations (measurements) of the gradients (1.11)-(1.12). The main ideas of this approach will be considered in this book.

1.4 Randomized strategies and equivalent smoothed problems

The link between the standard nonlinear programming and smoothed stochastic optimization problems will be stated.

For all $x \in X$, the nonconvex set Φ of the values of the functions $\phi_j(x)$, $(j = 0, ..., m)$ is defined as follows:

$$\Phi = \left\{ v \in R^{m+1} : v_j = \phi_j(x) \ (j = 0, ..., m), \ x \in X \right\} \tag{1.24}$$

A typical nonconvex set is depicted in Figure 1.1. In this Figure we have considered only one constraint. The coordinates of the point A are respectively the value of the cost function and the value of the considered constraint for a given value of x.

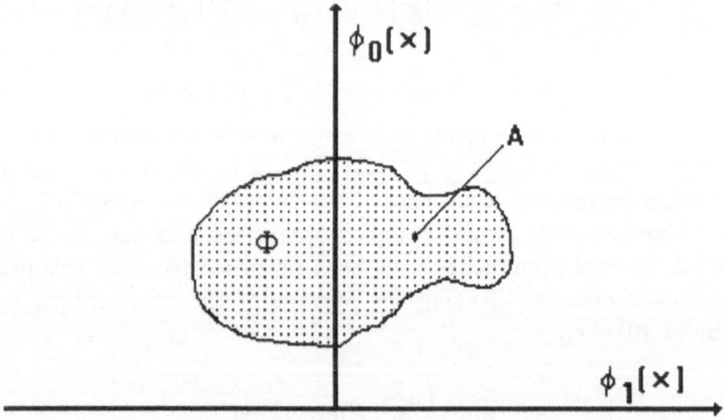

FIGURE 1.1. The set of possible values of $\phi_j(x)$, $(j = 0, ..., m)$.

For the standard nonlinear programming problem, the functions are assumed to be convex. The corresponding set Φ of all the possible values $(\phi_j(x))$ in the $(m+1)$ dimensional space is convex too, and as a consequence, many classical optimization techniques [16] including the stochastic approximation methods [13]-[14]-[15], can be used.

For nonconvex functions $\phi_j(x)$,$(j = 0, ..., m)$, the set Φ is nonconvex too. To use the standard optimization techniques for solving this problem which naturally is multimodal, it is necessary to solve several locally convex problems and to select the optimum among the obtained solutions.

Now, let us consider the stochastic nonlinear problem (1.16)-(1.17) where at least

- one of the functions $\phi_j(x)$ $(j = 0, ..., m)$ is nonconvex
- one of the functions $\phi_j(x)$ (or all of them) is nonsmooth, i.e., we must use only observations of zero-order (1.10).

To solve such nonlinear stochastic programming problem, which at first glance seems to be very complex, we will use the approach presented in [17]-[18]-[19].

Let us introduce the distribution function $F_x(z)$ of the vector $x = x(\omega) \in X$. It is assumed to be given on the same probability space $(\Omega, \mathcal{F}, P)$, and consider the following constrained optimization problem

$$\overline{\phi}_0(p) := \int_{x \in X} \phi_0(x)p(x)dx \to \inf_{p(x)} \tag{1.25}$$

$$\overline{\phi}_j(p) := \int_{x \in X} \phi_j(x)p(x)dx \leq 0 \tag{1.26}$$

where $p(x)$ is the corresponding probability distribution which is related to the distribution function F_x by

$$F_x(x) = \int_{-\infty}^{x_1} ... \int_{-\infty}^{x_M} p(v)dv \tag{1.27}$$

and satisfying the probability measure properties

$$p(x) \geq 0 \; \forall x \in X \subseteq R^M \tag{1.28}$$

$$\int_{-\infty}^{\infty} ... \int_{-\infty}^{\infty} p(v)dv = 1 \tag{1.29}$$

We will consider not only absolutely continuous distribution functions satisfying (1.27) but also other distribution functions like the $\delta-$functions.

The $\delta-$function operates as follows:

$$p(x) = \delta(x - x_0) \Longrightarrow \tag{1.30}$$

$$\int_{-\infty}^{\infty} ... \int_{-\infty}^{\infty} \psi(x)p(x)dx = \int_{-\infty}^{\infty} ... \int_{-\infty}^{\infty} \psi(x)\delta(x - x_0)dx = \psi(x_0) \tag{1.31}$$

for any function $\psi(x)$, $x \in R^M$.

Notice that, independently of the convexity properties of the initially defined functions $\phi_j(x)$ $(j = 0, ..., m)$, the smooth problem (1.25)-(1.26) subject to the probability measure constraints (1.28)-(1.29) is always a linear programming problem in the space of distributions $p(x)$ (including δ-functions (1.30)).

The relation between the standard nonlinear programming problem (1.16)-(1.17) and the corresponding smooth problem
(1.25)-(1.29) will be stated by the following theorem.

Theorem 1. If

1. *the initial standard stochastic nonlinear programming problem (1.16)-(1.17) has a solution x^* (not necessary unique), i.e., there exists a point $x^* \in X$ such that*

$$\begin{array}{c} \phi_0(x^*) \leq \phi_0(x^*) \qquad \forall x \in X \\ \phi_j(x^*) \leq 0, \ (j = 1, ..., m) \end{array} \tag{1.32}$$

2. *the constraints -(1.17) of the initial problem (1.16) satisfy the Slater's condition for $x = \overline{x} \in X$*
3. *the set X is compact (closed and bounded)*
4. *$\phi_j(x)$ $(j = 1, ..., m)$ are continuous functions on X*

Then, the corresponding smooth problem (1.25)-(1.29) has a solution too, *i.e.,*

$$\begin{array}{c} \exists p(x^*) : \overline{\phi}_0(p^*) \leq \phi_0(p) \ \forall p \in S_X^N \\ \overline{\phi}_j(p^*) \leq 0, \ (j = 1, ..., m) \end{array} \tag{1.33}$$

$$p, p^* \in S_X^N := \left\{ p(x) : p(x) \geq 0 \ \forall x \in X \subseteq R^M, \right. \tag{1.34}$$

$$\left. \int_{-\infty}^{\infty} ... \int_{-\infty}^{\infty} p(x) dx = 1 \right\}$$

and the minimal values $\phi_0(x^*)$ and $\overline{\phi}_0(p^*)$ of the cost functions $\phi_0(x^*)$ (1.32) and $\overline{\phi}_0(p)$ (1.33) are connected by the following relations:

$$\phi_0(x^*) - \overline{\phi}_0(p^*) = (1 - z_2^*)\phi_0(x^*) - (1 - z_1^*)\phi_0(\overline{x}) - (z_1^* - z_2^*)\phi_0(x^{**}) \tag{1.35}$$

where

$$x^{**} = \arg\min_{x \in X/X_0} \phi_0(x) \tag{1.36}$$

$$X_0 := \{x \in X : \phi_j(x) \leq 0 \ (j = 1, ..., m)\} \tag{1.37}$$

(z_1^*, z_2^*) *are the solution of the linear programming problem*

$$g_0(z_1, z_2) := \phi_0(\overline{x}) + z_1 [\phi_0(x^{**}) - \phi_0(\overline{x})] + z_2 [\phi_0(x^*) - \phi_0(x^{**})] \to \inf_{z_1, z_2}$$

$$g_j(z_1, z_2) := \phi_j(\overline{x}) + z_1 [\phi_j(x^{**}) - \phi_j(\overline{x})] + z_2 [\phi j(x^*) - \phi_j(x^{**})] \leq 0 \tag{1.38}$$

$$j = 1, ..., m; \ 0 \leq z_2^* \leq z_1^* \leq 1$$

and p^* *is equal to*

$$p^* = (1 - z_1^*)\delta(x - \overline{x}) + (z_1^* - z_2^*)\delta(x - x^{**}) + z_2^*\delta(x - x^*) \tag{1.39}$$

Proof.

Let us consider the following probability distribution representation

$$p(x) = (1 - \alpha)\delta(x - \overline{x}) + \alpha q(x), \ \alpha \in [0, 1] \tag{1.40}$$

where $\overline{x} \in X$ is the Slater's point (1.6) and $q(x)$ is any density function.

This formulation (1.40) represents some kind of parametrization of the class of all density functions (any density function $p(x)$ can be expressed as a function of any density function $q(x)$ and the parameter $\alpha \in [0, 1]$).

Using (1.40) it follows

$$\overline{\phi}_0(p) = (1 - \alpha)\phi_0(\overline{x}) + \alpha\overline{\phi}_0(q) = \tag{1.41}$$

$$= (1 - \alpha)\phi_0(\overline{x}) + \alpha \left[\int_{X_0} \phi_0(x)q(x)dx + \int_{X/X_0} \phi_0(x)q(x)dx \right]$$

As

$$\phi_0(x^*) \leq \phi_0(x) \ \forall x \in X_0 \tag{1.42}$$

From (1.41) we derive

$$\overline{\phi}_0(p) \geq (1 - \alpha)\phi_0(\overline{x}) + \alpha [\phi_0(x^*)q_0 + \phi_0(x^{**})(1 - q_0)] \tag{1.43}$$

with

$$q_0 := \mathbf{P}\{x \in X_0\} \tag{1.44}$$

Notice that x^{**} exists because the function $\phi_0(x)$ and the set X are assumed to be respectively continuous and compact.

The lower estimation (1.43) is sharp, i.e., there exists a density $p(x)$, verifying the identity in (1.43) and which is equal to

$$p(x) = (1-\alpha)\delta(x-\overline{x}) + \alpha\left[q_0\delta(x-x^*) + (1-q_0)\delta(x-x^{**})\right] \quad (1.45)$$

Substituting (1.45) into (1.26) we obtain

$$\overline{\phi}_j(p) = (1-\alpha)\overline{\phi}_j(\overline{x}) + \alpha\left[\phi_j(x^*)q_0 + \phi_j(x^{**})(1-q_0)\right] \leq 0 \quad (1.46)$$

Let us introduce two new variables

$$z_1 := \alpha \in [0,1] \text{ and } z_2 := \alpha q_0 \in [0,1]$$

Solving this linear programming problem (1.38) with respect to the variables z_1 and z_2 we obtain the minimum value $\overline{\phi}_0(p^*)$

$$\begin{gathered} \overline{\phi}_0(p^*) = \min_{z_1,z_2} g_0(z_1,z_2) = g_0(z_1^*,z_2^*) \\ g_j(z_1^*,z_2^*) \leq 0,\ 0 \leq z_2^* \leq z_1^* \leq 1 \end{gathered} \quad (1.47)$$

and hence p^* is given by (1.39). Theorem is proved. ■

Corollary 1. *For the unconstrained optimization problem, the minimal value of the cost functions associated respectively with the initial problem ((1.16)-(1.17)) and the smooth problem, coincide, i.e.,*

$$\phi_0(x^*) = \overline{\phi}_0(p^*),\ p^* = \delta(x-x^*) \quad (1.48)$$

Proof.

In this case $X_0 = X$, $X/X_0 = \emptyset$

It follows that $x^{**} = x^*$ and $\phi_0(x^*) = \phi_0(x^{**})$, and hence from (1.35) we obtain:

$$\phi_0(x^*) - \overline{\phi}_0(p^*) = (1-z_1^*)\left[\phi_0(x^*) - \phi_0(\overline{x})\right] \leq 0, \forall \overline{x} \in X \quad (1.49)$$

From (1.38) it follows that

$$g_0(z_1,z_2) = \phi_0(\overline{x}) + z_1\left[\phi_0(x^*) - \phi_0(\overline{x})\right] \to \inf_{\substack{z_1,z_2 \\ 0 \leq z_2 \leq z_1 \leq 1}}$$

has a solution for $z_1^* = 1$, from which and (1.49) we obtain (1.48) $\forall \overline{x} \in X$.

Corollary is proved. ■

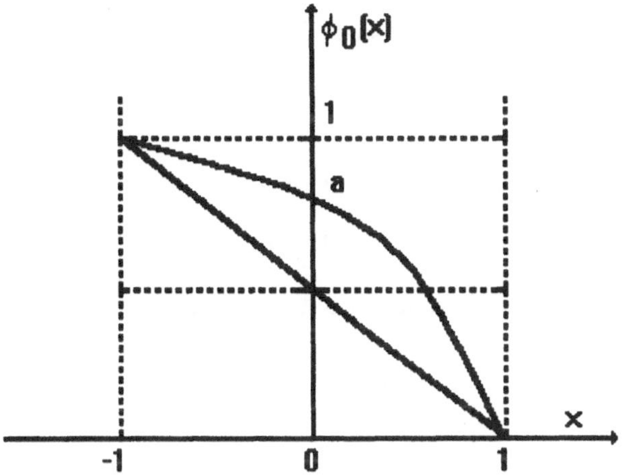

FIGURE 1.2. The cost function $\phi_0(x)$ defined on the set $[-1, 1]$.

Remark 3 *From (1.35) it follows that in general the minimal value of the cost function (1.25) of the smooth problem (1.25)-(1.26) can be less than the minimal value $\phi_0(x^*)$ of the cost function (1.16) of the deterministic problem (1.16)-(1.17). This is due to the fact that average constraints are less stronger than deterministic constraints.*

Consider the cost function $\phi_0(x)$ depicted in Figure 1.2.

The deterministic problem formulated as

$$\phi_0(x) \inf_{x \in X = [-1,1]}$$

subject to

$$-1 \leq x \leq 0$$

has a solution

$$x^* = 0, \phi_0(x^*) = a > \frac{1}{2}$$

The corresponding smooth problem can be formulated as follows:

$$\mathbf{E}\{\phi_0(x)\} \inf_{p(x)}$$

subject to

$$-1 \leq \mathbf{E}\{x\} \leq 0$$
$$p(x) = 0, \; x \notin X = [-1,1]$$

Taking $p(x)$ equal to

$$p(x) = \alpha\delta(x+1) + (1-\alpha)\delta(x-1) \tag{1.50}$$

we derive

$$-1 \leq \alpha(-1) + (1-\alpha) \leq 0$$

From which it follows that

$$\alpha \geq \frac{1}{2}$$

Calculating the cost function, we derive

$$\mathbf{E}\{\phi_0(x)\} = \alpha\phi_0(-1) + (1-\alpha)\phi_0(1) = \alpha\phi_0(-1)$$

Minimizing the right side toward α, we obtain

$$\mathbf{E}\{\phi_0(x)\} \geq \frac{\phi_0(-1)}{2} = \frac{1}{2}$$

So, for (1.50) we obtain

$$\mathbf{E}\{\phi_0(x)\} < \phi(x^*)$$

1.5 Caratheodory theorem and stochastic linear problem on finite set

Let us start with the definition of the envelope *en* X of a set X before introducing the Caratheodory theorem which will be useful in the following developments (the significance of any definition , of course, resides in its consequences and applications).

Definition 1 *The set* X *is said to be* ***the envelope of a set*** $X \in R^M$ *if for any point* $\overline{x} \in en\ X$ *there exist two points* $x_1, x_2 \in X$ *and a parameter* $\alpha \in [0,1]$ *such that*

$$\overline{x} = (1-\alpha)x_1 + \alpha x_2 \tag{1.51}$$

Let us recall the Caratheodory theorem from geometry theory (convex sets) [6].

Theorem 2. *(Caratheodory [6]). Any point* $\overline{x}$ *belonging to the envelope set* enX *of a set* $X \in R^M$ *can be expressed as a linear combination of only* $(M+1)$ *points* $x_s \in X$ $(s = 1, ..., M+1)$, *i.e.,* $\forall \overline{x} \in en\ X$ $\exists x_s \in X \ (s = 1, ..., M+1), \alpha \in S^{M+1} := \left\{ \alpha_s \geq 0, \sum_{s=1}^{M+1} \alpha_s = 1 \right\}$ *such that*

$$\overline{x} = \sum_{s=1}^{M+1} \alpha_s x_s$$

The proof of this theorem is given in [6].

The next theorem discusses several properties of this set X.

Theorem 3. *On the class of continuous functions $\phi_j(x)$ $(j = 0, ..., m)$ the following properties*

1. *hold:*

$$\overline{\Phi} = \left\{ v_j = \overline{\phi}_j(p) : p \in S_X^N \right\} \text{ is a convex set} \tag{1.52}$$

2.

$$\overline{\Phi} = en\ \Phi \tag{1.53}$$

Proof.

To prove (1.52), let us consider two distributions $p_1(x)$ and $p_2(x) \in S_X^M$ and construct the set of values

$$\overline{\phi}_j^{\alpha}(p_1, p_2) := (1-\alpha)\overline{\phi}_j(p_1) + \alpha\overline{\phi}_j(p_2), \ \ \alpha \in [0,1]$$

If we show that for any $\alpha \in [0,1]$ the point $\left\{\overline{\phi}_j^{\alpha}(p_1,p_2)(j = 0, ..., m)\right\}$ belongs to $\overline{\Phi}$ we state the first result of this theorem. Indeed,

$$\overline{\phi}_j^{\alpha}(p_1, p_2) = \int_{x \in X} \phi_j(x) \left[\alpha p_1(x) + (1-\alpha)p_2(x)\right] dx =$$

$$= \int_{x \in X} \phi_j(x) p_\alpha(x) dx$$

where

$$p_\alpha(x) := (1-\alpha)p_1(x) + \alpha p_2(x) \tag{1.54}$$

It is easy to show that $p_\alpha(x) \in S_X^M$. Hence, $p_\alpha(x)$ is a distribution which corresponds to the point $\left\{\overline{\phi}_j^{\alpha}(p_1,p_2)(j = 0, ..., m)\right\} \in \overline{\Phi}$. This result is true for any $\alpha \in [0,1]$ and any distributions $p_1(x)$, $p_2(x) \in S_X^M$. This complete the proof of statement (1.52).

To prove (1.53) let us notice that any point $V \in \Phi$ corresponds to the distribution

$$p(x) = \delta(x - x_0)$$

in (1.25), (1.26), i.e.,

$$\phi_j(x_0) = \overline{\phi}_j(\delta(x - x_0)), \ \ (j = 0, ..., m)$$

Hence, we have the following inclusion :

$$\Phi \subseteq \overline{\Phi}$$

According to Caratheodory theorem [6], any point $V \in en\ \Phi$ can be expressed in the following form

$$V = \begin{bmatrix} \overline{\phi}_0 \\ \overline{\phi}_1 \\ \cdot \\ \cdot \\ \cdot \\ \overline{\phi}_m \end{bmatrix} = \sum_{s=1}^{m+2} \alpha_s \begin{bmatrix} \phi_0(x_s) \\ \phi_1(x_s) \\ \cdot \\ \cdot \\ \cdot \\ \phi_m(x_s) \end{bmatrix}$$

$$\alpha \in S^{m+2} := \left\{ \alpha_s \geq 0, \ \sum_{s=1}^{m+2} \alpha_s = 1 \right\}$$

hence,

$$V = \begin{bmatrix} \overline{\phi}_0(\sum_{s=1}^{m+2} \alpha_s \delta(x - x_s)) \\ \cdot \\ \cdot \\ \cdot \\ \overline{\phi}_m(\sum_{s=1}^{m+2} \alpha_s \delta(x - x_s)) \end{bmatrix}$$

and, taking into account that the linear combination

$$\sum_{s=1}^{m+2} \alpha_s \delta(x - x_s)$$

of δ-functions with $\alpha \in S^{m+2}$ is a distribution, we derive that

$$\Phi \subseteq en\ \Phi \subseteq \overline{\Phi} \tag{1.55}$$

Let us consider any $p(x) \in S_X^M$. We have

$$\overline{\phi}_j(p) = \int_{x \in X} \phi_j(x) p(x) dx \tag{1.56}$$

Let us prove that any $p(x)$, corresponding to (1.56) can be expressed in the following form

$$p(x) = \sum_{s=1}^{m+2} \alpha_s \delta(x - x_s), \ x, x_s \in X, \ \alpha \in S^{m+2} \tag{1.57}$$

where the points $\{x_s\}$ satisfy the assumption of this theorem.

To prove (1.57), let us again use the Caratheodory theorem [6]. According to (1.52) the set $\overline{\Phi}$ is convex, then for any fixed distribution p the corresponding vector

$$\overline{\phi}^T(p) := (\overline{\phi}_0(p), ..., \overline{\phi}_m(p)) \in R^{m+1}$$

can be expressed as follows

$$\overline{\phi}(p) = \sum_{s=1}^{m+2} \alpha_s \overline{\phi}(p_s) \tag{1.58}$$

where p_s is a distribution function. Because of the continuity of the functions $\phi_j(x)$ there exist the points x_s such that

$$\overline{\phi}_j(p_s) = \phi_j(x_s) \ (j = 0, ..., m)$$

and, taking into account (1.56), we conclude that

$$\overline{\phi}_j(p_s) = \phi_j(x_s) = \int\limits_{x \in X} \phi_j(x)\delta(x - x_s)dx$$

and hence, from (1.58), we finally derive that

$$\overline{\phi}(p) = \sum_{s=1}^{m+2} \alpha_s \overline{\phi}(p_s) =$$

$$= \sum_{s=1}^{m+2} \alpha_s \int\limits_{x \in X} \phi(x)\delta(x - x_s)dx =$$

$$= \int\limits_{x \in X} \phi(x) \sum_{s=1}^{m+2} \alpha_s \delta(x - x_s)dx, \ \phi^T(x) :=$$

$$:= (\phi_0(x), ..., \phi_m(x)) \in R^{m+1}$$

So, the representation (1.57) holds. It means that

$$en \ \Phi \supseteq \overline{\Phi} \tag{1.59}$$

Combining (1.55) and (1.59) we finally obtain

$$en \ \Phi = \overline{\Phi}$$

Theorem is proved. ■

Remark 4 *The statement (1.59) can be also proved by using Kall's theorem [20] (see Appendix A)).*

Indeed, to prove the correctness of assertion (1.57), let us first state the similar representation for $(m+3)$ points (linear combination of nonminimal numbers of points), i.e.,

$$p(x) = \sum_{s=1}^{m+3} \alpha_s \delta(x - x_s), \ x, x_s \in X, \ \alpha \in S^{m+3} \tag{1.60}$$

We must prove that such $\alpha \in S^{m+3}$ and $x_s \in X$ exist. Substituting (1.60) into (1.56), we obtain

$$\overline{\phi}_j(p) = \sum_{s=1}^{m+3} \alpha_s \phi_j(x_s) \tag{1.61}$$

and eliminating α_{m+3} ($\alpha_{m+3} = 1 - \sum_{s=1}^{m+1} \alpha_s$) from (1.61), we obtain

$$\overline{\phi}_j(p) = \sum_{s=1}^{m+2} \alpha_s \phi_j(x_s) + \phi_j(x_{m+3})(1 - \sum_{s=1}^{m+2} \alpha_s)$$

or

$$\overline{\phi}_j(p) - \phi_j(x_{m+3}) = \sum_{s=1}^{m+2} \alpha_s \left[\phi_j(x_s) - \phi_j(x_{m+3})\right] \tag{1.62}$$

According to the continuity property of the functions $\phi_j(x)$ $(j = 0, ..., m)$ we can find the points $\{x_s\}_{s=1,...,m+1}$ such that

$$\det A_{m+1,m+1} \neq 0$$

where $A_{m+1,m+1}$ is the submatrix containing the first $(m+1)$-columns of the matrix $A_{m+1,m+2} = \|a_{js}\|, a_{js} := \phi_j(x_s) - \phi_j(x_{m+3}), j = 0, ..., m;$ $s = 1, ..., m+2$

Then, we can rewrite the last relation (1.62) in the following matrix form

$$b = A_{m+1,m+2} \cdot \alpha \tag{1.63}$$

where

$$\begin{aligned} b^T &= (b_0, ..., b_m), \ b_j := \overline{\phi}_j(p) - \phi_j(x_{m+3}) \\ A_{m+1,m+2} &= [a_{js}], \ a_{js} := \phi_j(x_s) - \phi_j(x_{m+3}) \\ &\quad (j = 0, ..., m; s = 1, ..., m+2) \\ \alpha^T &= (\alpha_1, ..., \alpha_{m+2}), \ \alpha_s \geq 0 \end{aligned}$$

Applying Kall's theorem [20] (see Appendix A) for the existence of the nonnegative vector α and the points $x_s (s = 1, ..., m+1)$ satisfying the linear equation (1.63), it is necessary and sufficient that there exist $\mu \geq 0$ and $\lambda_j < 0$ $(j = 1, ..., m+1)$ such that

$$\sum_{l=1}^{m+1} \lambda_l A_l = \mu A_{m+2}$$

where A_l is the l-column of the matrix $A_{m+1,m+2}$. This relation can be always verified if the points $x_s (s = 1, ..., m+3)$ are selected according to the following rule:

$$sign\,(\phi_j(x_l) - \phi_j(x_{m+3})) = -sign\,(\phi_j(x_{m+2}) - \phi_j(x_{m+3}))$$

$$\forall l = 1, ..., m+1$$

According to the Caratheodory theorem any linear combination of $(m+3)$ points can be represented as a linear combination of only $(m+2)$-points (1.61) . Hence the formulation (1.63) takes place.

The next important statement follows directly from this theorem.

Corollary 2. *For the class of continuous functions, the stochastic programming problem (1.25)-(1.26) is equivalent to the following non-linear programming problem*

$$\sum_{s=1}^{m+2} p(s)\phi_0(x_s) \to \inf_{p\in S^{m+2}, x_s \in X} \tag{1.64}$$

$$\sum_{s=1}^{m+2} p(s)\phi_j(x_s) \leq 0 \ (j = 1, ..., m)$$

Let us denote the corresponding minimal point by p^{**}, x_s^{**} $(s = 1, ..., m+2)$. This point may be not unique.

The most important consequence of this theorem is the following one: the optimization problem is entirely characterized by

- a set of $(m+2)$ vectors $x_k \in X$
- $(m+2)$ values of the probability vector p_k $\left(p_k \geq 0, \sum_{k=1}^{m+2} p_k = 1, (k = 1, ..., m+2)\right)$.

and can be reformulated as follows

$$\inf_{z \in Z} h_0(z) \tag{1.65}$$

subject to

$$h_j(z) \leq 0, \ (j = 1, ..., m) \tag{1.66}$$

where

$$z = \{x_0, ..., x_{m+1}, p_0, ..., p_{m+1}\} \tag{1.67}$$

$$z \in R^s, s = (m+2)(n+1)$$

$$x_k \in X,\ p_k \geq 0,\ \sum_{k=1}^{m+2} p_k = 1$$

and,

$$h_j(z) = \sum_{k=1}^{m+2} f_j(x_k)p_k,\ (j = 0, ..., m) \tag{1.68}$$

The next theorem shows that this nonlinear programming problem (1.64) given on the compact set X in the case of Lipschitzian functions $\phi_j(x)$ $(j = 0, ..., m)$ can be approximated by the corresponding linear stochastic problem on a finite set.

Theorem 4. *If*

1. *X is a compact set of diameter D which can be partitioned into a number of nonempty subsets X_k $(k = 1, ..., N)$ having no intersection, i.e.,*

$$X = \bigcup_{k=1}^{N} X_k,\ X_i \bigcap_{i \neq j} X_j = \emptyset \tag{1.69}$$

2. *the functions $\phi_j(x)$ $(j = 0, ..., m)$ are Lipschitzian on each subset X_k, i.e.,*

$$\left|\phi_j(x^{'}) - \phi_j(x^{''})\right| \leq L_k^j \left|x^{'} - x^{''}\right| \ \forall x^{'}, x^{''} \in X_k \tag{1.70}$$

then, there exist fixed points $\{x_k^*\}$ $(x_k^* \in X_k,\ k = 1, ..., N)$ and an enough large positive integer N such that the discrete distribution

$$\widetilde{p}(x) = \sum_{k=1}^{N} p^*(k)\delta(x - x_k^*)$$

with the vector $p^* \in S^N$ which is a solution of the linear stochastic programming problem

$$\sum_{k=1}^{N} p(k)\phi_0(x_k^*) \to \inf_{p \in S^N} \tag{1.71}$$

$$\sum_{k=1}^{N} p(k)\phi_j(x_k^*) \leq 0\ (j = 1, ..., m) \tag{1.72}$$

satisfies the constraints (1.72) with $\varepsilon-$accuracy, i.e.,

$$\overline{\phi}_j(\widetilde{p}) \leq \varepsilon_j\ (j = 1, ..., m) \tag{1.73}$$

and the corresponding loss function $\overline{\phi}_0(\widetilde{p})$ will deviate from the minimal value $\overline{\phi}_0(p^{**})$ of the initial programming problem (1.64) not more than ε, i.e.,

$$\left|\sum_{k=1}^{N} p^{**}(k)\phi_0(x_{s_k}^{**}) - \sum_{k=1}^{N} p^{*}(k)\phi_0(x_k^{*})\right| \leq \varepsilon_0 \tag{1.74}$$

where

$$\varepsilon_j \geq \frac{D \max\limits_{k=1,\dots,N} L_k^j}{N} \quad (j = 0, \dots, m) \tag{1.75}$$

Proof.

Let $p^{**}, x_s^{**} (s = 1, \dots, m+1)$ be the optimal point (solution) of the problem (1.71), (1.72) and the partition of the compact X is realized such that each subset X_k could contain not more than one of the extremal points x_s^{**}. Hence, for any $(j = 0, \dots, m)$ we obtain

$$\sum_{k=1}^{N} p(k)\phi_j(x_k^{*}) = \sum_{k=1}^{N} p(k) \left[\phi_j(x_k^{*}) - \right.$$

$$\left. - \sum_{s=1}^{m+2} \phi_j(x_s^{**})\chi(x_k^{*}, x_s^{**} \in X_k)\right] +$$

$$+ \sum_{k=1}^{N} p(k) \sum_{s=1}^{m+2} \phi_j(x_s^{**})\chi(x_k^{*}, x_s^{**} \in X_k)$$

Let us denote by $x_{s_k}^{**}$ the optimal point $x_s^{**} \in X_k$. Then

$$\sum_{s=1}^{m+2} \phi_j(x_s^{**})\chi(x_k^{*}, x_s^{**} \in X_k) = \phi_j(x_{s_k}^{**})$$

and hence, for any $p \in S^N$

$$\sum_{k=1}^{N} p(k)\phi_j(x_k^{*}) = \sum_{k=1}^{N} p(k) \left[\phi_j(x_k^{*}) - \phi_j(x_{s_k}^{**})\right] + \tag{1.76}$$

$$+ \sum_{k=1}^{N} p(k)\phi_j(x_s^{**})$$

For $(j = 1, \dots, m)$ and for any $p \in S^N$, satisfying the constraints (1.72), we have

$$\sum_{k=1}^{N} p(k)\phi_j(x_s^{**}) \leq \left|\sum_{k=1}^{N} p(k) \left[\phi_j(x_k^{*}) - \phi_j(x_{s_k}^{**})\right]\right| \leq$$

$$\leq \sum_{k=1}^{N} p(k) \left| \phi_j(x_k^*) - \phi_j(x_{s_k}^{**}) \right| \leq$$

$$\leq \sum_{k=1}^{N} p(k) L_k^j \left\| x_k^* - x_{s_k}^{**} \right\| \leq$$

$$\leq \sum_{k=1}^{N} p(k) L_k^j \sup_{x,y \in X_k} \|x - y\| \leq$$

$$\leq \sum_{k=1}^{N} p(k) L_k^j d_k$$

Selecting $d_k := \sup_{x,y \in X_k} \|x - y\| \leq D/N$ we derive

$$\sum_{k=1}^{N} p(k) \phi_j(x_s^{**}) \leq \frac{D}{N} \max_{k=1,\dots,N} L_k^j := \varepsilon_j \ (j = 1, \dots, m) \tag{1.77}$$

Using now the fact that the point p^* minimizes the left side of (1.76) for $j = 0$,we derive:

$$\sum_{k=1}^{N} p^*(k) \phi_j(x_k^*) \leq \sum_{k=1}^{N} p(k) \phi_j(x_k^*) =$$

$$= \sum_{k=1}^{N} p(k) \left[\phi_j(x_k^*) - \phi_j(x_{s_k}^{**}) \right] + \sum_{k=1}^{N} p(k) \phi_j(x_s^{**})$$

Then, for $p(k) = p^{**}(k)$ we finally derive

$$\left| \sum_{k=1}^{N} p^{**}(k) \phi_0(x_{s_k}^{**}) - \sum_{k=1}^{N} p^*(k) \phi_0(x_k^*) \right| \leq$$

$$\leq \left| \sum_{k=1}^{N} p^{**}(k) \left[\phi_0(x_k^*) - \phi_0(x_{s_k}^{**}) \right] \right| \leq$$

$$\leq \sum_{k=1}^{N} p^{**}(k) \left| \phi_0(x_k^*) - \phi_0(x_{s_k}^{**}) \right| \leq \sum_{k=1}^{N} p^{**}(k) L_k^0 \left\| x_k^* - x_{s_k}^{**} \right\| \leq$$

$$\leq \sum_{k=1}^{N} p^{**}(k) L_k^0 \sup_{x,y \in X_k} \|x - y\| \leq$$

$$\leq \sum_{k=1}^{N} p^{**}(k) L_k^0 d_k$$

Selecting $d_k \leq D/N$ we obtain

$$\left|\sum_{k=1}^{N} p^{**}(k)\phi_0(x^{**}_{s_k}) - \sum_{k=1}^{N} p^{*}(k)\phi_0(x^{*}_{k})\right| \leq D/N \max_{k=1,...,N} L^0_k$$

We can satisfy (1.74) with an ε−accuracy if we choose

$$D/N \max_{k=1,...,N} L^0_k \leq \varepsilon_0$$

Theorem is proved. ■

Corollary 3. *Under the assumptions of this theorem, to obtain an ε−approximation of the initial nonlinear programming problem (1.64) it is enough to use the partition of the given compact set X with the diameter D into a subsets X_k $(k = 1, ..., N)$ with diameters*

$$d_k \leq \frac{D}{N} \leq \frac{\varepsilon}{\max_{k=1,...,N} L^0_k}$$

and the integer N, characterizing the number of such subsets, must satisfy the following inequality:

$$N \geq \frac{D \max_{k=1,...,N} L^0_k}{\varepsilon} \tag{1.78}$$

This corollary gives a means of determining the accuracy of the approximation of optimization problem on continuous set by optimization problems on finite (discrete) set.

In summary, Let us notice that a discrete optimization problem can be formulated as the behaviour of a learning automaton with finite set of actions in a random environment corresponding to the optimization problem to be solved. Indeed, the discrete set related to the optimization problem to be solved can be associated to the set of control actions of a learning automaton operating in a random environment. The loss functions related to this environment are equal to the values of the optimized function.

1.6 Conclusion

The main conclusions are the following:

- The linear stochastic optimization problem can arise directly (if it is initially formulated on a discrete set) or as an approximation with an ε-accuracy of the stochastic optimization problem on a continuous set.

- Independently on the properties (convexity, differentiability, etc.) of the functions, involved in a nonlinear programming problem, the ε-equivalent linear stochastic optimization problem on discrete set is a well-defined problem and represents a linear programming problem in the space of distributions (multidimensional simplex).

- The formulation of the stochastic optimization problem in the form (1.71) allows us to apply successfully learning automata for its solution.

- In summary, this chapter shows how to tie the stochastic optimization problems to the behaviour of learning automata operating in random environment (stationary, nonstationary, single, multi-teacher environments).

References

[1] Bertsekas D P 1982 *Constrained Optimization and Lagrange Multiplier Methods.* Academic Press, New York

[2] Rockafellar R T 1993 Lagrange Multipliers and Optimality. *SIAM Review* 35:183-238

[3] Najim K, Oppenheim G 1991 Learning Systems: Theory and Applications. *IEE Proceedings Computer and Digital Techniques* 138:183-192

[4] Najim K, Poznyak A S 1994 *Learning Automata: Theory and Applications.* Pergamon Press, Oxford

[5] Zangwill W I 1969 *Nonlinear Programming: A Unified Approach.* Prentice-Hall, Englewood Cliffs

[6] Rockafellar R T 1970 *Convex Analysis.* Princeton University Press, Princeton:

[7] Spingarn J E, Rockafellar R T 1979 The genetic nature of optimal conditions in nonlinear programs. *Mathematics of Operation Research* 4:425-430

[8] W.I. Zangwill W I, Garcia C B 1981 *PATWAYS to Solutions, Fixed Points, and Equilibria,* Prentice-Hall, Englewood Cliffs

[9] Charnes A, Cooper W W 1959 Chance constrained programming. *Man. Sci.* 6:73-79

[10] Vajda S 1972 *Probabilistic Programming.* Academic Press, New York

[11] Ash B B 1972 *Real Analysis and Probability.* Academic Press, New York

[12] Doob J L 1953 *Stochastic Processes.* John Wiley & Sons,New York

[13] Kiefer J, Wolfowitz J 1952 Stochastic estimation of the maximum of a regression function. *Ann. Math. Stat.* 23:462-466

[14] Kushner H J, Clark D S 1978 *Stochastic Approximation Methods for Constrained and Unconstrained Systems.* Springer Verlag, Berlin

[15] Gelfand S B, Mitter S K 1991 Recursive stochastic algorithms for global optimization in R^d. *SIAM J. Control and Optimization* 29:999-1018

[16] Polak E 1971 *Computational Methods in Optimization.* Academic Press, New York

[17] Kaplinskii A I, Propoi A I 1970 Stochastic approach to nonlinear programming problems. *Automation and Remote Control* 31:448-459

[18] Kaplinskii A I, Poznyak A S, Propoi A I 1971 Optimality conditions for certain stochastic programming problems. *Automation and Remote Control* 32:1210-1218

[19] Kaplinskii A I, Poznyak A S, Propoi A I 1971 Some methods for the solution of stochastic programming problems. *Automation and Remote Control* 32:1609-1616

[20] Kall P 1966 Qualitative aussagen zu einegen problem der stochastichen programmierung. *Z. Wahrsch. verw. Geb* 6:246-272

2

On Learning Automata

2.1 Introduction

Among the most attractive areas of research in the field of control engineering are adaptive and learning systems [1]-[2]-[3]-[4]-[5]-[6]-[7]-[8]-[9]. Learning systems have been shown to be an efficient tool to deal with a large number of engineering problems [9]. They are information processing systems whose architecture and behaviour are inspired by the structure of biological systems (the organism is born with relatively little initial knowledge and learns actions that are appropriate through trial and error) [5]-[10]. The vocabulary and the concepts associated with learning automata are borrowed from biology and psychology. The experiments performed by Skinner simply illustrate the behaviour of a Reward/Inaction learning automaton. A pigeon was placed in a cage with a red disk mounted on one of the walls. If by chance, the pigeon pecks at the red disk, it receives a certain amount of grain whereas if it pecks elsewhere, it receives no reward. It is not long before the pigeon associates the pecking of the red disk with being rewarded with food. There are also a myriad of other examples where this simple kind of learning is evidenced by the modification of behaviour [11].

Adaptive control methods [2]-[3]-[4] generally address the problem of control of systems with unknown parameters but known structure. In adaptive control strategies, the behaviour of the system is slightly improved at every sampling period by estimating in real-time the parameters (model or control law parameters) to attain the desired control objective. In learning automata [6], the probability distribution (p_n) is recursively updated to optimize some learning goal. Learning automata have attracted considerable interest in the last decades due to their potential usefulness in a variety of engineering problems which are characterized by nonlinearity and a high level of uncertainty [12]. In fact, later developments in stochastic control theory took into account uncertainties that might be present in a given process; stochastic control was effected by assuming that the probabilistic characteristics of the uncertainties are known. Frequently, the uncertainties are of a higher order, and even the probabilistic characteristics such as the distribution functions may not be completely known. It is then necessary to learn (acquire) additional information [13].

Broadly speaking, learning automata can be classified into three categories: deterministic, fixed-structure and variable-structure. In the deterministic automata, the transition and the output matrices are deterministic. A fixed-structure automaton is one whose transition and output functions are time invariant. Variable-structure stochastic automata possess transition and output functions which evolve as the learning process proceeds [6]-[8]. The automaton structure changes as the system learns the information. The use of variable-structure stochastic automata leads to a reduction of states in comparison with deterministic automata. A learning system interacts with an environment and learns the optimal action which the environment offers. It bases its decision on the information gained by selecting actions, seeing if the choice is rewarded, and then updating its probability distribution. This cycle continues until the learning process is terminated. The environment is characterized by its penalty probabilities. An environment is said to be stationary if its penalty probabilities are constant. Otherwise it is nonstationary.

Notice that the term "learning" is used by several authors in neural networks synthesis an in association with stochastic approximation techniques [9]-[13]-[14]-[15].

The reference [9] presents several theoretical results concerning the convergence and the estimation of the convergence rate and discusses in fair detail some of the applications of learning automata. Most of the available studies relate to the behaviour of learning automata in stationary environments [6]-[7]-[8]-[9]. The problem concerning the behaviour of learning automata in nonstationary environments is difficult, and few results are known [6]-[7]-[16]-[17]-[18]-[19]-[20]-[21]-[22]. Baba and Mogami [22] have shown that an extended form of the scheme proposed by Thathachar and Ramakrishan [23] ensures absolute expediency in a nonstationary environment having the property that there exists a unique path which receives the least sum of the penalty strengths in the sense of mathematical expectation. Poznyak and Najim [24] have studied the behaviour of learning automata in asymptotically stationary environments. In this study, several theoretical results were stated. These results concern the properties of reinforcement schemes, normalized environment response and the optimal behaviour of different learning automata. A nonstationary environment arises indirectly in connection with hierarchical system of learning automata [6]-[23]. It has been shown in [9] that the use of hierarchical system of learning automata accelerates the learning process. The latter reference also discusses in fair detail some of the applications of hierarchical structure of learning automata. The conventional automaton model of a learning system consists of a stochastic automaton operating in a single environment. Nevertheless, an automaton can be made to operate in more than one random medium [25]-[26]. Baba [7] has introduced the concept of average weighted reward and several norms (expediency, ε-optimality) of the learning behaviours of stochastic automata in multi-teacher environment. He has also proposed a class of reinforcement schemes which directly use the responses from the multi-teacher environment (the number of rewards).

A learning system is a sequential machine characterized by a set of actions, a probability distribution and a reinforcement scheme. An extensive literature has been dedicated to the behaviour of learning automata with fixed action set [8]. The concept of the behaviour of automata where the number of the actions available at each time is time-varying has been studied by Thathachar and coauthors [27]-[28]. In this study [27], convergence results have been stated for binary environment responses (P-model environment). An important aspect of convergence not considered is the rate of convergence which concerns the speed of operation of the automaton [27]. Learning automata with changing number of actions are relevant in the modelling of several problems (CPU job scheduling, optimal path in stochastic networks, etc.). Learning automaton with continuous input (S-model environment) where the number of automaton actions are changed in real-time have been considered by Poznyak and Najim [28]. Learning automata with continuous inputs and changing number of actions have been used for optimization purposes [9]-[28].

The analysis and the statement of the convergence properties of learning systems are usually derived using the Lyapunov approach and the martingale theory (see Appendix A) [29]-[30]-[31]-[32].

2.2 Learning automaton

A learning automaton may be considered as a system which modifies its control strategy on the basis of its experience in order to reach good control (optimization) performances in spite of unpredictable changes in the environment where it operates. In other words, learning automata should, by collecting and processing current information regarding the environment, be capable of changing their structure and parameters as time evolves to achieve the desired goal or the optimal performance (in some sense). An automaton is an adaptive discrete machine described by:

$$\{\Xi, U, \mathcal{R}, \{\xi_n\}, \{u_n\}, \{p_n\}, T\}$$

where:

(i) Ξ is the automaton input bounded set.

(ii) U denotes the set $\{u(1), u(2), ..., u(N)\}$ of actions of the automaton.

(iii) $\mathcal{R} = (\ \Omega, \mathcal{F}, \mathbf{P}\ )$ a probability space.

(iv) $\{\xi_n\}$ is a sequence of automaton inputs (environment response, $\xi_n \in \Xi$) provided by the environment in a binary (P-model environment) or continuous (S-model environment) form.

(v) $\{u_n\}$ is a sequence of automaton outputs (actions).

(vi) $p_n = [p_n(1), p_n(2), ..., p_n(N)]^T$ is the probability distribution at time n

$$p_n(i) = \mathbf{P}\{\omega : u_n = u(i) \ / \ \mathcal{F}_{n-1}\} \text{ and } \sum_{i=1}^{N} p_n(i) = 1 \ , \forall n$$

where $\mathcal{F}_n = \sigma(\xi_1, u_1, p_1; ...; \xi_n, u_n, p_n)$ is the σ-algebra generated by the corresponding events $(\mathcal{F}_n \in \mathcal{F})$.

(vii) $c_n = [c_n(1), c_n(2), ..., c_n(N)]^T$ is the conditional mathematical expectation vector of the environment responses (at time n).

(viii) T represents the reinforcement scheme (updating scheme) which changes the probability vector p_n to p_{n+1}:

$$p_{n+1} = p_n + \gamma_n T_n(p_n; \{\xi_t\}_{t=1,...,n}; \{u_t\}_{t=1,...,n}) \tag{2.1}$$

$$p_1(i) > 0 \quad \forall i = 1, ..., N$$

where γ_n is a scalar correction factor and the vector $T_n(.) = \left[T_n^1(.), ..., T_n^N(.)\right]^T$ satisfies the following conditions (for preserving probability measure):

$$\sum_{i=1}^{N} T_n^i(.) = 0, \forall n \tag{2.2}$$

$$p_n(i) + \gamma_n T_n^i(.) \in [0, 1] \quad \forall n, \forall i = 1, ..., N \tag{2.3}$$

This is the heart of the learning automaton. Different reinforcement schemes have been proposed in the literature. A reinforcement scheme can be linear or nonlinear. Sometimes it is advantageous to update p_n according to different schemes depending on the intervals in which the value of p_n lies.

The loss function Φ_n associated with the learning automaton is given by

$$\Phi_n = \frac{1}{n} \sum_{t=1}^{n} \xi_t \tag{2.4}$$

It is a useful quantity for judging the behaviour of a learning automaton.

A learning system is a stochastic automaton connected in a feedback loop with a random environment as shown in Figure 2.1.

The description of environments and the learning automata will be given in the next sections.

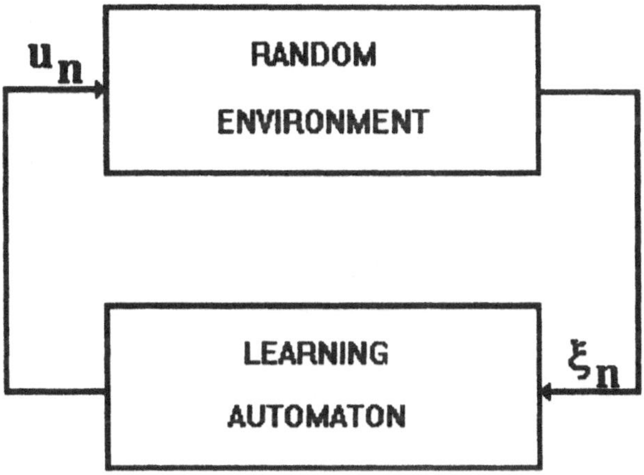

FIGURE 2.1. Automaton-environment interaction.

2.3 Environment

Learning systems are adaptive machines which interact with an environment and which dynamically learn the optimal action which the environment offers. The role of the environment (medium) is to establish the relation between the actions of the automaton and the signals received at its input. The environment (or medium) is a term that can cover just about anything. It includes all the external conditions and influences [6]. The environment produces random responses whose statistics depend on the current stimulus or input. The environment offers the automaton a finite set of actions. The automaton is constrained to choose one of these actions. The environment is said to be P-model environment if its response belongs to the set $\{0, 1\}$ (binary input). It is said to be an S-model environment when its response takes an arbitrary value in the closed segment $[0, 1]$ (continuous response). In Q-model environment, the environment output can assume a finite number of values in the interval $[0, 1]$. In automatic control, the environment corresponds to the process to be controlled or optimized (to introduce control signal in a system means to couple the system to its environment). The environment is characterized by its penalty probabilities c_i. An environment is said to be stationary if its penalty probabilities are constant. Otherwise it is nonstationary. Having considered single environment, we are ready for our next topic: multi-environments (multi-teacher). The remainder of this section is devoted to multi-teacher environments [7]. A typical learning automaton operating in a multi-teacher environment, is depicted in Figure 2.2. It consists of a set of "teachers". Each teachers reacts to the automaton response. These reactions (responses) at time

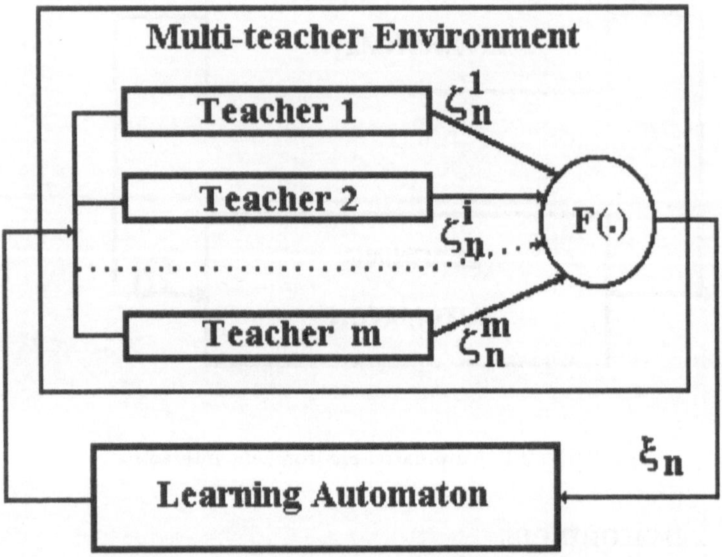

FIGURE 2.2. Multi-teacher environment.

n, are denoted by ζ_n^j $(j = 1, ..., m)$. They can be binary (P-model environment) or continuous (S-model environment). The automaton input $\xi_n = F(\zeta_n^j, ..., \zeta_n^j)$ is constructed on the basis of these responses. Some examples of the transformation $F(\cdot)$ will be presented in Chapter 4. The attention given to this class of environments is due to their very interesting intrinsic properties. In fact they provide simple means to represents some engineering problem. For example, in process control, each teacher can be associated with a state of the process to be controlled. In constrained optimization problems, one teacher can be associated with the cost function and the other to the different constraints.

2.4 Reinforcement schemes

Reinforcement schemes were originally proposed in an attempt to model animal learning [33] and have since found successful application in the field of learning automata. A reinforcement scheme can be compared to the recursive estimation procedure used in adaptive control. The reinforcement scheme (learning or updating algorithm) generates p_{n+1} from p_n. Several algorithms for adjusting the probabilities after each sampling period (interaction with the environment) have been proposed [8]-[9]. They are based on incremental changes in the probabilities. The most commonly used employs a linear updating algorithm have been proposed by Bush and Mosteller [33]. All the reinforcement schemes described in the literature can be considered as being solutions of optimization problems [34]. Let us introduce the following average penalty function

$$J = \sum_{i=1}^{N} \{\Phi_i(p_n)E_i[1-\xi_i] - \Psi_{ii}(p_n)E_i[\xi_i]\} \tag{2.5}$$

where the functions $\Phi_i(p_n)$ and $\Psi_i(p_n)$ represent respectively the amount of change in the probability vector under reward ($\xi_i = 0$) and penalty ($\xi_i = 1$). ξ_i ($i = 1, 2$) which correspond to the reward and penalty environment responses. The interest here is in minimizing (2.5). According to Kiefer-Wolfowitz approximation method [35], the reinforcement scheme which minimizes the function J by setting the gradient of J equal to zero is derived.

If the selected action at time n is $u(i)$, the following algorithm is obtained:

$$p_{n+1}(i) = p_n(i) + \frac{\gamma_n}{p_n(i)}\left[\frac{\partial \Phi_i(p_n)}{\partial p_n(i)}(1-\xi_i) - \frac{\partial \Psi_i(p_n)}{\partial p_n(i)}\xi_i\right] \tag{2.6}$$

$$p_{n+1}(j) = p_n(j) - \tag{2.7}$$

$$-\frac{\gamma_n}{p_n(i)\,[N-1]}\left[\frac{\partial \Phi_i(p_n)}{\partial p_n(i)}(1-\xi_i) - \frac{\partial \Psi_i(p_n)}{\partial p_n(i)}\xi_i\right]$$

$$i \neq j$$

We have just derived one of the central result of reinforcement schemes. In general, the existing learning schemes differ in structure (linear, nonlinear, etc.) but they fall into the general framework (2.6-2.8). The functions associated with Bush-Mosteller [33], Shapiro-Narendra [36] and Varshavskii-Vorontsova [37] reinforcement schemes are given in Table 2.1.

Table 2.1. Different reinforcement schemes and their corresponding functions $\Phi(\cdot)$ and $\Psi(\cdot)$.

Autors	Function $\Phi(p_n)$	Function $\Psi(p_n)$
Bush and Mosteller	$p_n(i) - \frac{1}{2}p_n(i)^2$	$p_n(i)^2$
Varshavskii and	$cte - \frac{1}{2}p_n(i)^2 - \frac{1}{3}p_n(i)^3$	$\Phi_i(\cdot)$
Vorontsova		
Shapiro and Narendra	$p_n(i) - \frac{1}{2}p_n(i)^2$	0

The bush-Mosteller and Shapiro-Narendra are linear reinforcement schemes, while the Varshavskii-Vorontsova scheme is nonlinear. These schemes are relatively too general and should be efficiently adapted for use in specific environments (S-model, asymptotically stationary environments, etc.). A more general method for constructing learning schemes has been presented.

When the number of actions increases, the behaviour of the automaton will be slow and the computer memory capacity required for the implementation of the learning systems will also increase. These problems can be avoided by using a hierarchical structure of automata.

2.5 Hierarchical structure of learning automata

A hierarchical structure of learning automata is depicted in Figure 2.3. It is composed at different levels of single automata with limited number of actions [9]-[22]-[23]-[38]. The first level of the hierarchy comprises a single automaton with N actions. The second level is composed of N single automata (of N actions each) and the kth level is formed by N^{k-1} automata. The last level of the hierarchy contains N^N single automata. The hierarchical structure of automata operates as follows. The first level chooses randomly an action (say $u(1)$). This activates the corresponding automaton in the second level (each action of the automaton contained in the first level, is associated with an automaton belonging to the second level) which select an action (say $u(2/3)$). This in turn activates an automaton in the third level of the hierarchy. At the last level the action selected ($u(3/2_{/2})$) interacts with the environment (environment input).

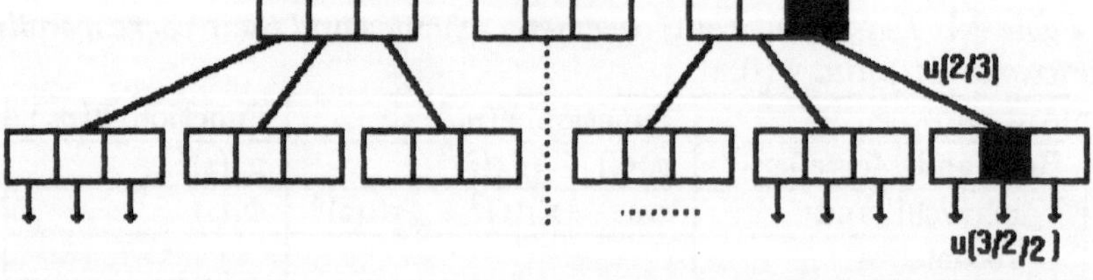

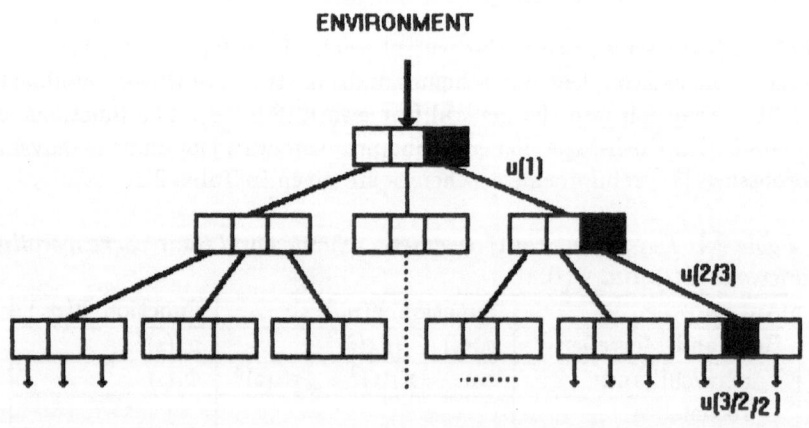

FIGURE 2.3. Hierarchical structure of learning automata.

The hierarchical structure of learning automata can be used for several purposes [9]-[38] (optimization, multiobjective analysis, etc.).

The probability measure

$$\sum_{i=1}^{N} p_n(i) = 1$$

is not ensured when the environment response do not belong to the interval $[0;1]$. it is then necessary to introduce some procedure for preserving the probability measure. Two solutions are commonly used: projection and normalization [9]-[34].

The next section concerns these procedures for environment response transformation and probability "scaling".

2.6 Normalization and projection

In this section, our main emphasis will be on normalization and projection procedures. When the probability do not belong to the probability variation domain $[0,1]$, it is necessary to use some procedure (operator or transformation) which guarantee that the probability measure remains satisfied at each time. The normalization procedure [39]-[40] process the environment output (response) while the projection acts on the probabilities.

2.6.1 NORMALIZATION PROCEDURE

The normalization procedure has been introduced by Najim and Poznyak [9] and used in the context of learning automata with continuous input (S-model environment) for multimodal functions optimization. We shall present a very simple normalization procedure. Let us consider the multi-teacher environment represented in Figure 2.2, and assume that the response of each teacher is binary, i.e.,

$$\zeta_n^j \in \{0,1\}, \ (j = 1, ..., m)$$

The normalized environment response $\tilde{\xi}_n$ (automaton input) can be constructed as follows:

$$\tilde{\xi}_n = \frac{\sum_{j=1}^{m} \zeta_n^j}{m} \in [0,1] \tag{2.8}$$

This normalization procedure ensures that the automaton input belongs to the unit segment $[0,1]$.

A more general methods for constructing normalized environment responses will be introduced and used in the next chapters.

In what follows, the projection operator is described.

2.6.2 PROJECTION ALGORITHM

The projection procedure can be applied for binary and non-binary environment responses [34]. It is a useful procedure when the probabilities will not range over the interval [0, 1]. The projection operator Π onto the simplex S is defined as

$$\Pi(z) = \begin{cases} z \; if \; z \in S \\ z^* \; if \; z \notin S, \; \| z - z^* \| = \min_{y \in S} \| z - y \| \end{cases}$$

The projection operator $\Pi(\cdot)$ is characterized by the following property:

$$\| p_n - \Pi(q) \| \leq \| p_n - q \|$$

The N-dimensional simplex S is defined by:

$$S \equiv \left\{ p_n : \sum_{i=1}^{N} p_n(i) = 1, \; p_n(i) \geq 0 \right\}, \quad i = 1, ..., N$$

It consists of:
$C_N^1 = N$ vertices Γ_1^j
$C_N^2 = \frac{1}{2} N(N-1)$ edges (two dimensional faces) Γ_2^j
.
.
$C_N^{N-1} = N(N-1)$ dimensional faces Γ_{N-1}^j

The point for which the coordinates are all equal to zero (except the jth which equals 1) is the vertex Γ_1^j. The face Γ_m^j $(m \geq 2)$ is the subset

$$\Gamma_m^j = \{P : X \in D_m, \; p(i) \geq 0\}$$

of one of the hyperplane D_n defined as follows

$$D_n = p_n : \sum_{i=1}^{N} p_n(i) = 1, \; p_n(i) \geq 0, \quad i = 1, ..., N$$

The projection of p_n is defined as follows:

$$\Pi(p_n) = p_n^* : \| p_n - p_n^* \| = \min_{y \in S} \| p_n^* - y \| \tag{2.9}$$

It is obvious that $p_n^* \in \Gamma_k^j$ for a certain k.

Note that finding $p_n^* = \Pi(p_n)$ is equivalent to finding the point on the simplex S which is closest to the projection $p_n(D_n)$ of the point p_n onto D_n.

From the definition (2.9), it follows that

$$\min_{y \in S} \| y - p_n^* \| = \min_{y \in S} \| (y - p_n(D_n)) + (p_n(D_n) - p_n^*) \| =$$

$$= \| (p_n(D_n) - p_n^*) \|^2 + \| (y - p_n(D_n)) \|^2$$

The following lemma gives the tool for calculating the projection $\Pi(p_n)$ of p_n.

Lemma 1. *The face Γ_{N-1} closest to the point $p_n(D_n)$ has an orthogonal vector*

$$a_{N-1} = (\underbrace{1, 1, ..., 1, 0}_{j}, ..., 1) \tag{2.10}$$

The index j corresponds to the smallest component of the vector (point) $p_n(D_n)$, i.e.,

$$j = \left\{ i : (p_n(D_n))_j = \min_i (p_n(D_n))_i \right\} \tag{2.11}$$

Proof.

The distance V between a given point $z = (z_1, z_2, ...) \in D_n$ and the vertex

$$\Gamma_1^k = \left\{ \underbrace{0, 0, ..., 0, 1}_{k}, 0, ..., 0 \right\}$$

is equal to

$$V^2(z, \Gamma_1^k) = (z_k - 1)^2 + \sum_{i \neq k}^{N} (z_i)^2 = \sum_{i=1}^{N} (z_i)^2 + (1 - 2z_k)$$

Then,

$$V^2(z, \Gamma_1^k) - V^2(z, \Gamma_1^l) = 2(z_l - z_k)$$

i.e., The most distant vertex corresponds to the smallest component

$$p_n(k) = \min_l p_n(l)$$

Consequently, the face Γ_{N-1} which lies opposite to the vertex and is closest to the point z has an orthogonal vector a_{N-1} (2.10). ■

The projection of $p_n(D_m)$ onto the hyperplane $p_n(D_m)$ $(1 < m \leq N)$ is accomplished according to

$$(p_n(D_m))_j = \left(p_n(j) + \frac{1 + \sum_{i=1}^{N} a(i) p_n(j)}{m} \right) a(j), \ (j = 1, ..., m) \tag{2.12}$$

where $a(j)$ are the components of the vector which is orthogonal to the hyperplane D_m.

We note that the projection procedure is accomplished in a number of steps not exceeding N.

The projection onto the simplex S_ε defined by

$$S_\varepsilon \equiv \left\{ p_n : \sum_{i=1}^{N} p_n(i) = 1,\ p_n(i) > \varepsilon \right\}, \quad i = 1, ..., N$$

can be carried out by means of the change of variables

$$p_n(i) = p_n^*(i) + \varepsilon$$

The structure of the projection operator algorithm is indicated below.

If $p_n \notin S$ then,

1. find $p_n(D_m)$ according to (2.12) for $m = N$
2. If $p_n(D_m) \notin S$, then find the smallest component of the vector which is orthogonal to the closest face Γ_1^k (2.11).

The geometrical interpretation of the projection procedure for the two-dimensional case $[p_n(1)\ p_n(2)]^T$ is depicted in Figure 2.4.

Five cases $(P(i), i = a, b, c, d, e)$ have been considered. The coordinates of each point P are the probabilities $p_n(1)$ and $p_n(2)$. The thin lines represent the behaviour of the projection operator. We obtain the following mapping

$$P(a) \xrightarrow{\Pi} A,\ P(b) \xrightarrow{\Pi} B,\ P(c) \xrightarrow{\Pi} C,\ P(d) \xrightarrow{\Pi} C,\ P(e) \xrightarrow{\Pi} C$$

For example, the coordinates (x_a, y_a) of the point A correspond to the projection of $P(a)$. Referring to Figure 2.3, the projection operator can considered as the map that associates with any point $P(.)$ of the plane a point belonging to the segment BC.

Normalization and projection have many nice properties. There are useful in the context of optimization and give the user a greater flexibility. The disadvantage of projection is that it is time consuming. In other words, the use of the projection operator results in a relatively sluggish behaviour of the learning system.

To deal with numerical problems which are due to round-off errors, the probability vector can be scaled by:

$$s_p = \sum_{i=1}^{N} p_n(i)$$

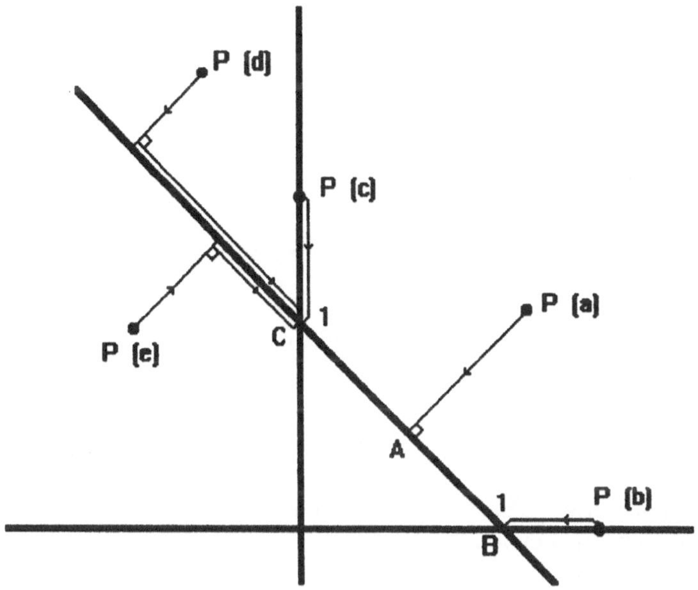

FIGURE 2.4. Graphical interpretation of the projection procedure.

2.7 Conclusion

This Chapter has been concerned with an introduction to learning automata. Several notions and definitions have been given. Both single and hierarchical structure of learning automata were defined. It has been shown that several reinforcement schemes can be associated with the minimization of some functional. The normalization procedure and the projection operator have been introduced for preserving the probability measure. Many of the results presented in this chapter are basic for theoretical background and the behaviour of learning automata in different kinds of random environments (stationary, nonstationary, etc.). In the next chapters it will be shown that learning automata can be used to solve a large class of stochastic optimization problems and that there exhibit a satisfactory degree of robustness (insensitivity to uncertainties, etc.).

References

[1] Wiener N 1948 *Cybernetics*. The Technology Press John Wiley, New York

[2] Bellman R 1973 *Adaptive Control Processes-A Guided Tour*. Princeton University Press, Princeton

[3] K.J. Åström K J, Wittenmark B 1984 *Computer Controlled Systems Theory and Design*. Prentice-Hall, Englewood Cliffs

[4] Najim K 1988 *Control of Liquid-Liquid Extraction Columns*. Gordon and Breach, London

[5] Tsetlin M L 1973 *Automaton Theory and Modeling of Biological Systems*. Academic Press, New York

[6] Narendra K S, Thathachar M A L 1989 *Learning Automata an Introduction*. Prentice-Hall, Englewood Cliffs

[7] Baba N 1984 *New Topics in Learning Automata Theory and Applications*. Springer-Verlag, Berlin

[8] Najim K, Oppenheim G 1991 Learning systems: theory and applications. *IEE Proceedings Computer and Digital Techniques* 138:183-192

[9] Najim K, Poznyak A S 1994 *Learning Automata: Theory and Applications*. Pergamon Press, Oxford

[10] Walter W G 1953 *The Living Brain*. Norton, New York

[11] Caudill M, Butler C 1990 *Naturally Intelligent Systems*. MIT Press, Cambridge

[12] Najim K 1989 *Process Modeling and Control in Chemical Engineering*. Marcel Dekker, New York

[13] Tsypkin Ya Z 1971 *Adaptation and Learning in Automatic Systems*. Academic Press, New York

[14] Pathak-Pal A, Pal S K 1987 Learning with mislabeled training samples using stochastic approximation. *IEEE Transactions on Systems, Man, and Cybernetics* 17:1072-1077

[15] Zurada J M 1992 *Artificial Neural Systems*. West Publishing Company, New York

[16] Srikantakumar P R, Narendra K S 1982 A learning model for routing in telephone networks. *SIAM J. Control and Optimization* 20:34-57

[17] Barto A, Anandan P, Anderson C W 1986 Cooperatively in networks of pattern recognizing stochastic learning automata. In: Narendra K S (ed) 1986 *Adaptive and Learning Systems : Theory and Applications*. Plenum Press, New York

[18] Narendra K S, Viswanathan R 1972 A two-level system of stochastic automata for periodic random environments. *IEEE Trans. Syst. Man, and Cybern.* 2:285-289

[19] Nedzelnitsky O V Jr, Narendra K S 1987 Nonstationary models of learning automata routing in data communication networks. *IEEE Trans. Syst. Man, and Cybern.* 17:1004-1015

[20] Narendra K S, Thathachar M A L 1980 On the behavior of learning automata in a changing environment with application to telephone traffic routine. *IEEE Trans. Syst. Man, and Cybern.* 10:262-269

[21] Koditschek D E, Narendra K S 1977 Fixed structure automata in a multi-teacher environment. *IEEE Trans. Syst. Man, and Cybern.* 7:616-624

[22] Baba N, Mogami Y 1988 Learning behaviours of hierarchical structure of stochastic automata in a nonstationary multi-teacher environment. *Int. J. Systems Sci.* 19:1345-1350

[23] Thathachar M A L, Ramakrishnan K R 1981 A hierarchical system of learning automata. *IEEE Trans. Syst. Man, and Cybern.* 11:236-241

[24] Poznyak A S, Najim K 1997 On the behaviour of learning automata in nonstationary environments. to appear in *European Journal of Control*

[25] Baba N 1983 The absolutely expedient nonlinear reinforcement schemes under the unknown multiteacher environment. *IEEE Trans. Syst. Man, and Cybern.* 13:100-107

[26] Baba N 1983 On the learning behaviours of variable-structure stochastic automaton in the general N-teacher environment. *IEEE Trans. Syst. Man, and Cybern.* 13:224-231

[27] Thathachar M A L, Harita B R 1987 Learning automata with changing number of actions. *IEEE Trans. Syst. Man, and Cybern.* 17:1095-1100

[28] Najim K, Poznyak A S 1996 Multimodal searching technique based on learning automata with continuous input and changing number of actions. *IEEE Trans. on Systems, Man, and Cybernetics* 26:666-673

[29] Doob J L 1953 *Stochastic Processes.* John Wiley & Sons, New York

[30] Robbins H, Siegmund D 1971 A convergence theorem for nonnegative almost supermartingales and some applications. In: Rustagi J S (ed) 1971 *Optimizing Methods in Statistics.* Academic Press, New York

[31] Nazin A V, Poznyak A S 1986 *Adaptive Choice of Variants.* (in Russian) Nauka, Moscow

[32] Neveu J 1975 *Discrete-Parameter Martingales.* North-Holland Publishing, Amsterdam

[33] Bush R R,Mosteller F 1958 *Stochastic Models for Learning.* John Wiley & Sons, New York

[34] Poznyak A S 1975 Investigation of the convergence of algorithms for the functioning of learning stochastic automata. *Automation and Remote Control* 36:77-91

[35] Kiefer J, Wolfowitz J 1952 Stochastic estimation of the maximum of a regression function," *Ann. Math. Stat.* 23:462-466

[36] Shapiro I J, Narendra K S 1969 Use of stochastic automata for parameter self optimization with multimodal performance criteria. *IEEE Trans. Syst. Man, and Cybern.* 5:352-361

[37] Varshavskii V I, Vorontsova I P 1963 On the behavior of stochastic automata with variable structure. *Automation and Remote Control* 24:327-333

[38] Narendra K S, Parthasarathy K 1991 Learning automata approach to hierarchical multiobjective analysis. *IEEE Trans. Syst. Man, and Cybern.* 21:263-273

[39] Poznyak A S, Najim K, Chtourou M 1996 Learning automata with continuous inputs and their application for multimodal functions optimization. *Int. J. of Systems Science* 27:87-95

[40] Poznyak A S, Najim K, Chtourou M 1996 Analysis of the behaviour of multilevel hierarchical systems of learning automata and their application for multimodal functions optimization. *Int. J. of Systems Science* 27:97-112

3

Unconstrained Optimization Problems

3.1 Introduction

Several engineering problems require a multimodal functions optimization strategy. Usually, the function $f(x)$ to be optimized is not explicitly known; only samples disturbed values $f(x)$ at various settings of x can be observed, making the usual numerical optimization procedures useless.

Random search techniques [1]-[2], the model trust region technique [3], simulated annealing [4] and learning automata [5]-[6] have been widely used for the optimization of functions where more than one local optimum exists

Random search techniques are generally based on random sampling and search region contraction [1] or on stochastic approximation techniques [7]-[8]-[9]. In the model trust region technique [3], the step for a new iterate is obtained by minimizing a local quadratic function over a restricted spherical region centered on the current iterate.

The simulated annealing method is suitable for the optimization of large scale systems and multimodal functions [4] and is based on the principles of thermodynamics involving the way liquids freeze and crystallize. Learning systems have made a significant impact on many areas of engineering problems including modelling, control, optimization, pattern recognition, signal processing, neural networks synthesis,fuzzy logic processor training and diagnosis. They are attractive and provide interesting methods for solving complex nonlinear problems characterized by a high level of uncertainty [10]-[11]-[12]-[13]-[14]. The learning system consists of a learning automaton operating in a random environment (the problem to be solved). This chapter deals with the use of learning automata with continuous input (S-model environment) for solving unconstrained stochastic optimization problems (optimization of functions with nonunique stationary points where the observations are disturbed by random variables, etc.).

This chapter presents an unified analysis of the commonly used reinforcement schemes. It is shown that these unconstrained stochastic optimization problems which are given on compact sets, are equivalent with an $\varepsilon-$accuracy to stochastic optimization problems given on finite sets.

The first part of this chapter will be concerned with learning automata with fixed number of actions while the second part will be dedicated to learning automata where the number of automaton-actions changes with time [15]-[16]-[17].

3.2 Statement of the Stochastic Optimization Problem

Let us consider the function $f(x)$ which is a real-valued function of a vector parameter $x \in X$ where X is a compact in R^M. We would like to find the value $x = x^*$ which minimizes this function, i.e.

$$x^* = \arg \min_{x \in X \subset R^m} f(x) \tag{3.1}$$

There are almost no conditions concerning the function $f(x)$ (continuity, unimodality, differentiability, convexity, etc.) to be optimized. We are concerned with a global optimization problem of multimodal and nondifferentiable functions.

Let y_n be the observation of the function $f(x)$ at the point $x_n \in X$, i.e.,

$$y_n = f(x_n) + w_n \tag{3.2}$$

where w_n is the observation noise (disturbance) at time n.

The stochastic optimization problem on the given compact which we intend to address is: **using the observations $\{y_n\}$, construct the sequence $\{x_n\}$ which converges (in some probability sense) to the optimal point x^*.**

Consider now a quantification $\{X_i\}$ of the admissible compact region $X \subset R^M$:

$$X_i \subset X,\ X_j \underset{i \neq j}{\cap} X_i = \emptyset\,(i, j = 1, ..., N)\,,\ \bigcup_{i=1}^{N} X_i = X \subset R^M \tag{3.3}$$

where $x_n \in \{x(1), x(2), ..., x(N)\} := \mathcal{X}$, $x(i) \in X_i$. Here the points $x(i)$ are some fixed point (for example, the center point of the corresponding subset X_i) and $w_n = w_n(\omega)$ is a random variable given on a probability space $(\Omega, \mathcal{F}, \mathbf{P})$($\omega \in \Omega$ - a space of elementary events) and characterizing the observation noise associated with the point x_n.

The next lemma states the connection between the original stochastic optimization problem on the given compact X, formulated above, and the corresponding stochastic optimization problem on the finite set $\mathcal{X}$.

Lemma 1. *Let us assume that on each subset X_i the optimized function $f(x)$ is Lipschitzian, i.e., there exists a constant L_i such that for any x' and $x'' \in X_i$ the following inequality is fulfilled*

$$\left|f(x') - f(x'')\right| \leq L_i \left\|x' - x''\right\|$$

Then, for any $\varepsilon > 0$ there exists a quantification $\{X_i\}$ $(i = 1, ..., N)$ of the admissible compact region $X \subset R^M$ such that the minimal values of the optimized function $f(x)$ on the compact X and on the discrete set $\mathcal{X}$ differ not more than ε. Moreover, for this purpose, the quantification number N must satisfy the following inequality

$$N \geq \frac{D}{\varepsilon} \max_{i=1,...,N} L_i \tag{3.4}$$

where $D := \sup_{x,y \in X} \|x - y\|$ is the diameter of the given compact X.

Proof.

Let us denote by $x(\alpha)$ the point on the discrete set $\mathcal{X}$ where the minimal value of the optimized function is reached, i.e.,

$$x(\alpha) := \arg\min_{x \in \mathcal{X}} f(x) \tag{3.5}$$

Then, using the Lipschitzian property we directly derive

$$f(x(\alpha)) - f(x^*) = f(x(\alpha)) - \min_{x \in X} f(x) =$$

$$= f(x(\alpha)) - \min_{x \in X} \sum_{i=1}^{N} f(x)\chi(x \in X_i) =$$

$$= \min_{x \in X} \sum_{i=1}^{N} [f(x(\alpha)) - f(x)]\, \chi(x \in X_i) \leq$$

$$\leq \min_{x \in X} \sum_{i=1}^{N} [f(x(i)) - f(x)]\, \chi(x \in X_i) \leq$$

$$\leq \max_{x \in X} \sum_{i=1}^{N} |f(x(i)) - f(x)|\, \chi(x \in X_i) \leq \max_{x \in X} \sum_{i=1}^{N} L_i \, \|x(i) - x\|\, \chi(x \in X_i) \leq$$

$$\leq \max_{i=1,...,N} L_i d_i$$

Here $d_i := \sup_{x,y \in X_i} \|x - y\|$ is the diameter of the subset X_i. Taking into account that we can organize the quantification such that

$$d_i \leq \max_{i=1,...,N} d_i \leq \frac{D}{N}$$

we can always select N satisfying the condition

$$\max_{i=1,...,N} L_i d_i \leq \frac{D}{N} \max_{i=1,...,N} L_i \leq \varepsilon$$

Lemma is proved. ■

Based on this lemma we can formulate the following problem statement : **using the observations $\{y_n\}$ construct the sequence $\{x_n\}$ $(x_n \in \mathcal{X})$ which converges (in some probability sense) to the optimal point** $x(\alpha)$ (3.5).

The solution of this stochastic optimization problem, formulated on the discrete set $\mathcal{X}$ leads with any ε-accuracy to the solution of the initial stochastic optimization problem given on the compact X where the accuracy level ε is related to the quantification number N by inequality (3.4). The increase of the quantification number N leads to the decrease of the accuracy (increase of ε).

This optimization problem formulated on the discrete set $\mathcal{X}$ will be stated and solved as the behaviour of a learning automaton operating in a random environment. As a learning automaton is a sequential machine, the quantification number N will be associated with the number of actions of the learning automaton.

The next section deals with the link between the previous stochastic optimization problem and learning systems.

3.3 Learning Automata Design

A learning automaton is connected in a feedback loop to the random medium (environment), where the input to one is the output to the other.

Let us consider the reinforcement schemes (updating schemes) which are the mechanisms used to change the probability vector p_n to p_{n+1}:

a) *Nonprojectional Reinforcement Schemes*

$$p_{n+1} = p_n + \gamma_n T_n(p_n; \{\xi_t\}_{t=1,...,n}; \{u_t\}_{t=1,...,n}) \tag{3.6}$$

$$p_1(i) > 0 \quad \forall i = 1, ..., N$$

where γ_n is a scalar correction factor, the vector $T_n(.) = \left[T_n^1(.), ..., T_n^N(.)\right]^T$ satisfies the following conditions (for preserving probability measure):

$$\sum_{i=1}^{N} T_n^i(.) = 0, \forall n \tag{3.7}$$

$$p_n(i) + \gamma_n T_n^i(.) \in [0, 1] \qquad \forall n, \forall i = 1, ..., N \tag{3.8}$$

and $\{\xi_t\}_{t=1,...,n}$ is the sequence of environment responses $\xi_n \in R^1$ which in our case will be constructed on the basis of the available data (observations), i.e.,

$$\xi_n = \xi_n\left(\{y_t\}_{t=1,...,n}\right)$$

For the nonprojectional schemes it is usually assumed that the environment response belongs to the unit closed line segment, i.e.,

$$\xi_n \in [0,1]$$

b) *Projectional Reinforcement Schemes*

$$p_{n+1} = \pi_{\varepsilon_n}\left\{p_n + \gamma_n T_n(p_n; \{\xi_t\}_{t=1,...,n}; \{u_t\}_{t=1,...,n})\right\} \tag{3.9}$$

$$p_1(i) > 0 \quad \forall i = 1, ..., N$$

where $\pi_{\varepsilon_n}\{.\}$ is the projection operator onto the ε_n−simplex $S^N_{\varepsilon_n}$ defined as follows

$$S^N_{\varepsilon_n} := \left\{p = (p_1, ..., p_N) : p(i) \geq \varepsilon_n \geq 0, \ \sum_{i=1}^{N} p(i) = 1\right\} \tag{3.10}$$

(for these schemes conditions (3.7) and (3.8) are not obligatory).

Different reinforcement schemes, satisfying conditions (3.7) and (3.8), are given in Table 1 of the previous chapter (see also [14]).

In this study we shall be concerned with Bush-Mosteller [18], Shapiro-Narendra [19] and Varshavskii-Vorontsova [20] reinforcement schemes.

The loss function Φ_n (see chapter 2) associated with the learning automaton is usually given by

$$\Phi_n = \frac{1}{n}\sum_{t=1}^{n} \xi_t \tag{3.11}$$

It is a useful quantity for judging the behaviour of a learning automaton. We will show in the sequel that if a stochastic automaton minimizes its loss function (find the best control action $x(\alpha)$) then it automatically solves the corresponding unconstrained stochastic optimization problem on a discrete set.

Let us consider now **a random stationary environment** which responses ξ_n are characterized by the following two properties:

(H1) The conditional mathematical expectations of the environment responses exist and are stationary, i.e.,

$$\mathbf{E}\left\{\xi_n/\mathcal{F}_{n-1} \bigwedge u_n = u(i)\right\} = c(i), \quad \forall\, i = 1, ..., N \tag{3.12}$$

(H2) The conditional variances of the environment responses are bounded, i.e.,

$$\begin{aligned} \mathbf{E}\left\{(\xi_n - c_n(i))^2/\mathcal{F}_{n-1} \bigwedge u_n = u(i)\right\} &= \sigma^2(i), \forall\, i = 1, ..., N \\ \max_i \sigma^2(i) &:= \sigma^2 < \infty \end{aligned} \tag{3.13}$$

Here $\mathcal{F}_n \subset \mathcal{F}$ is the σ-algebra generated by the corresponding process, i.e.,

$$\mathcal{F}_n := \sigma\left(\xi_1, p_1, u_1; ...; \xi_n, p_n, u_n\right)$$

The next theorem states the equivalence between the problem related to the asymptotic minimization of the loss function (3.4) and the corresponding linear programming problem.

Theorem 1. *The problem of "Optimization of the Learning Automata Behaviour", which consists of finding a sequence $\{u_n\}$ of automata control actions minimizing asymptotically the loss function (3.4), i.e.*

$$\limsup_{n\to\infty} \Phi_n \overset{a.s.}{\to} \inf_{\{u_n\}} \tag{3.14}$$

under hypotheses (H1), (H2) coincides with the following linear programming problem

$$\sum_{i=1}^{N} c(i)p(i) \to \min_{p\in S^N} \tag{3.15}$$

in the sense that their minimal values coincide:

$$\inf_{\{u_n\}} \limsup_{n\to\infty} \Phi_n \overset{a.s.}{=} \min_{p\in S^N} \sum_{i=1}^{N} c(i)p(i) = \min_{i=1,...,N} c(i) := c(\alpha) \tag{3.16}$$

Proof.

It is clear that the minimal value of the function $\sum_{i=1}^{N} c(i)p(i)$ is equal to $c(\alpha)$ and can be achieved with the optimal pure strategy p^* defined as follows

$$p^*(i) = \delta_{i\alpha} \ (i = 1, ..., N) \tag{3.17}$$

Indeed,

$$\sum_{i=1}^{N} c(i)p(i) \geq \min_{p\in S^N} \sum_{i=1}^{N} c(i)p(i) \geq \min_{p\in S^N} \sum_{i=1}^{N} \min_{i=1,...,N} c(i)p(i) =$$

$$= c(\alpha) \min_{p\in S^N} \sum_{i=1}^{N} p(i) = c(\alpha)$$

and

$$c(\alpha) = \sum_{i=1}^{N} c(i)p^*(i)$$

To prove the first equality in (3.16), let us rewrite Φ_n (3.4) in the following form:

$$\Phi_n = \sum_{i=1}^{N} f_n(i)\bar{\xi}_n(i) \tag{3.18}$$

where the sequences $\{f_n(i)\}$ and $\{\bar{\xi}_n(i)\}$ are defined as follows

$$f_n(i) := \frac{1}{n}\sum_{t=1}^{n} \chi(u_t = u(i))$$

$$\bar{\xi}_n(i) := \begin{cases} \dfrac{\sum_{t=1}^{n} \xi_t(u(i),\omega)\chi(u_t=u(i))}{\sum_{t=1}^{n} \chi(u_t=u(i))} & \text{if } \sum_{t=1}^{n} \chi(u_t = u(i)) > 0 \\ 0 & \text{if } \sum_{t=1}^{n} \chi(u_t = u(i)) = 0 \end{cases}$$

Taking into account assumption (**H1**), and according to lemma A.12 [14], for almost all

$$\omega \in \mathcal{B}_i := \left\{\omega : \sum_{t=1}^{n} \chi(u_t = u(i)) = \infty\right\}$$

we have

$$\lim_{n\to\infty} \bar{\xi}_n(i) = c(i)$$

For almost all $\omega \notin \mathcal{B}_i$ we evidently have

$$\limsup_{n\to\infty} \left|\bar{\xi}_n(i)\right| < \infty$$

and, hence,

$$\lim_{n\to\infty} f_n(i) = 0$$

But the vector f_n which components are $f_n(i)$, belongs to the simplex S^N, and as a result any partial limit Φ of a the sequence $\{\Phi_n\}$ may be expressed, with probability 1, in the following form:

$$\Phi = \sum_{i=1}^{N} p(i)c(i), \ p \in S^N$$

where p is a partial limit of the sequence $\{f_n\}$. Hence,

$$\limsup_{n\to\infty} \Phi_n \geq \liminf_{n\to\infty} \Phi_n \overset{a.s.}{\geq} \min_{p\in S^N} \sum_{i=1}^{N} p(i)c(i)$$

and this inequality turn to an equality if

$$p_n(i) := \mathbf{P}\{u_n = u(i)|\mathcal{F}_{n-1}\} = p^*(i), \ (i = 1, ..., N)$$

Theorem is proved. ■

The next theorem is fundamental. It shows that if a nonstationary strategy $\{p_n\}$ which converges to the optimal strategy p^* is used, then the sequence $\{\Phi_n\}$ of loss functions reaches its minimal value $c(\alpha)$.

Theorem 2. *If under assumptions (H1), (H2) a reinforcement scheme generates a sequence* $\{p_n\}$, *which converges with probability one to the optimal strategy* p^* *(3.17), then, the sequences* $\{\Phi_n\}$ *of loss functions reaches its minimal value* $c(\alpha)$ *with probability one, i.e. if*

$$\|p_n - p^*\| = o_\omega(1) \underset{n\to\infty}{\overset{a.s.}{\to}} 0$$

then,

$$\lim_{n\to\infty} \Phi_n \overset{a.s.}{=} c(\alpha) \tag{3.19}$$

Proof.

In view of the strong law of large numbers [21]-[22] (assumptions under consideration), we have

$$\frac{1}{n}\sum_{t=1}^{n} \xi_t - \frac{1}{n}\sum_{t=1}^{n} E\{\xi_t|\mathcal{F}_{n-1}\} \underset{n\to\infty}{\overset{a.s.}{\to}} 0$$

But

$$\begin{aligned}
\frac{1}{n}\sum_{t=1}^{n} E\{\xi_t|\mathcal{F}_{n-1}\} &= \frac{1}{n}\sum_{t=1}^{n} c(i)p_n(i) = \\
&= \frac{1}{n}\sum_{t=1}^{n} c(i)p^*(i) + \frac{1}{n}\sum_{t=1}^{n} c(i)\left[p_n(i) - p^*(i)\right] = \\
&= c(\alpha) + o_\omega(1) \underset{n\to\infty}{\overset{a.s.}{\to}} c(\alpha)
\end{aligned}$$

Theorem is proved. ■

Corollary 1. *If there exists a continuous function* $W(p_n)$, *which on the trajectories* $\{p_n\}$ *possesses the following properties:*

1. $W(p) \geq W(p^*) = 0 \; \forall p \in S^N$
2. $W(p_n) \underset{n\to\infty}{\overset{a.s.}{\to}} 0$

then, this trajectory $\{p_n\}$ *is asymptotically optimal, i.e.,*

$$p_n \underset{n\to\infty}{\overset{a.s.}{\to}} p^*$$

Proof.

follows immediately from the property of continuity and the considered assumptions. ■

This scalar function $W(p)$ is called *Lyapunov function.* It will be used for the convergence analysis (some interesting properties) of different reinforcement schemes realizing an asymptotically optimal behaviour of the considered class of Learning Stochastic Automata. Notice that the main obstacle to using the Lyapunov approach is finding a suitable Lyapunov function.

3.4 Nonprojectional Reinforcement schemes

Let us now consider some nonprojectional learning algorithms (reinforcement schemes) of the type (3.6), and the following Lyapunov function:

$$W_n := \frac{1 - p(\alpha)}{p(\alpha)} \tag{3.20}$$

It is obvious that this function satisfies the first condition of the previous corollary. We would like to check the validity of the second condition for several commonly used reinforcement schemes applied to nonbinary environment responses, and to show then how to use them for optimization purposes.

3.4.1 Bush-Mosteller reinforcement scheme

The Bush-Mosteller scheme [18]-[14] is described by the following recurrent equation:

$$p_{n+1} = p_n + \gamma_n \left[e(u_n) - p_n + \xi_n (e^N - Ne(u_n))/(N-1) \right] \tag{3.21}$$

where

(i) $\xi(n) \in \Xi = [0,1]$,

(ii) $e^N = (1, ..., 1)^T \in R^N$,

(iii) $e(u(n)) = (0, 0, ..., 0, 1, 0, ..., 0)^T \in R^N$ (the ith component of this vector is equal to 1 if $u(n) = x_i$ and the other components are equal to zero),

(iv) $\gamma(n) \in (0,1)$.

The key to being able to analyse the behaviour of a learning automaton using the Bush-Mosteller reinforcement scheme is the following theorem.

Theorem 3. *If*

1. *the environment response sequence* $\{\xi_n\}$ *satisfies assumptions (H1),(H2)*
2. *the optimal action* $u_n = u(\alpha)$ *is unique, i.e.,*

$$\min_{i \neq \alpha} c(i) > c(\alpha) = 0$$

3. *the correction factor is selected as follows:*

$$\gamma_n = \frac{\gamma}{n+a}, \ \gamma \in (0,1), a > \gamma$$

then, the Lyapunov function (3.20) possesses the property

$$W(p_n) \underset{n \to \infty}{\overset{a.s.}{\to}} 0$$

and, hence, the reinforcement scheme (3.21) generates asymptotically an optimal sequence $\{p_n\}$*, such that*

$$p_n \underset{n \to \infty}{\overset{a.s.}{\to}} p^*$$

Proof.
Let us consider the following estimation:

$$p_{n+1}(i) = p_n(i) + \gamma_n[\chi(u_n = u(i)) - p_n(i) +$$

$$+\xi_n \left[1 - N\chi(u_n = u(i))\right]/(N-1)] = p_n(i)\ (1-\gamma_n) +$$

$$+\frac{\gamma_n}{N-1}\left\{\xi_n + \chi(u_n = u(i))\left[N(1-\xi_n) - 1\right]\right\} \geq \qquad (3.22)$$

$$\geq p_n(i)\ (1-\gamma_n) \geq \cdots \geq p_1(i) \prod_{t=1}^{n} (1-\gamma_t) > 0$$

From lemma A.4 [14], it follows that

$$\prod_{t=1}^{n}(1-\gamma_t) \geq \begin{cases} \left(\frac{a-\gamma}{n+a}\right)^{\gamma}, & a > \gamma \in (0,1) \\ \frac{a}{n+a}, & \gamma = 1,\ a > 0 \end{cases} \qquad (3.23)$$

The reinforcement scheme (3.21) leads to

$$\mathbf{E}\left\{\frac{1-p_{n+1}(\alpha)}{p_{n+1}(\alpha)}/\mathcal{F}_n\right\} \overset{a.s.}{=}$$

$$\overset{a.s.}{=} \sum_{i=1}^{N} \mathbf{E}\left\{\frac{1}{p_{n+1}(\alpha)} - 1/\mathcal{F}_{n-1} \bigwedge u_n = u(i)\right\} p_n(i) \overset{a.s.}{=}$$

$$= \sum_{i=1}^{N} \left(\frac{1}{\mathbf{E}\left\{p_{n+1}(\alpha)/\mathcal{F}_{n-1} \bigwedge u_n = u(i)\right\}} - 1\right) p_n(i) + s_n \tag{3.24}$$

where

$$s_n = \sum_{i=1}^{N} \mathbf{E}\left\{\frac{1}{p_{n+1}(\alpha)} - \right.$$

$$\left. - \frac{1}{\mathbf{E}\left\{p_{n+1}(\alpha)/\mathcal{F}_{n-1} \bigwedge u_n = u(i)\right\}}/\mathcal{F}_{n-1} \bigwedge u_n = u(i)\right\} p_n(i) \tag{3.25}$$

Let us now estimate the term s_n (3.25):

$$s_n = \sum_{i=1}^{N} \mathbf{E}\left\{ Y_n/\mathcal{F}_{n-1} \bigwedge u_n = u(i)\right\} p_n(i) =$$

$$= \sum_{i\neq\alpha}^{N} \mathbf{E}\left\{ Y_n/\mathcal{F}_{n-1} \bigwedge u_n = u(i)\right\} p_n(i) + \tag{3.26}$$

$$+\mathbf{E}\left\{\frac{\mathbf{E}\left\{p_{n+1}(\alpha)/\mathcal{F}_{n-1} \bigwedge u_n = u(\alpha)\right\} - p_{n+1}(\alpha)}{p_{n+1}(\alpha)\mathbf{E}\left\{p_{n+1}(\alpha)/\mathcal{F}_{n-1} \bigwedge u_n = u(i)\right\}}/\mathcal{F}_{n-1} \bigwedge u_n = u(\alpha)\right\} \times$$

$$\times p_n(\alpha)$$

where

$$Y_n = \frac{\mathbf{E}\left\{p_{n+1}(\alpha)/\mathcal{F}_{n-1} \bigwedge u_n = u(i)\right\} - p_{n+1}(\alpha)}{p_{n+1}(\alpha)\mathbf{E}\left\{p_{n+1}(\alpha)/\mathcal{F}_{n-1} \bigwedge u_n = u(i)\right\}}$$

Taking into account that

$$p_{n+1}(\alpha) = p_n(\alpha) + \gamma_n[\delta_{i,\alpha} - p_n(i) + \xi_n\left[1 - N\delta_{i,\alpha}\right]/(N-1)] =$$

$$= \begin{cases} p_n(\alpha)(1-\gamma_n) + \gamma_n & \text{if} \quad u(i) = u(\alpha) \ (\xi_n \overset{a.s.}{=} 0) \\ p_n(\alpha)(1-\gamma_n) + \gamma_n\xi_n/(N-1) & \text{if} \quad u(i) \neq u(\alpha) \end{cases}$$

we derive

$$\mathbf{E}\left\{ Y_n/\mathcal{F}_{n-1} \bigwedge u_n = u(\alpha)\right\} p_n(\alpha) \overset{a.s.}{=} 0$$

and from (3.26) we obtain

$$s_n = \sum_{i\neq\alpha}^{N} \mathbf{E}\left\{ Y_n/\mathcal{F}_{n-1} \bigwedge u_n = u(i) \right\} p_n(i) \overset{a.s.}{=}$$

$$= \gamma_{nN} \sum_{i\neq\alpha}^{N} \mathbf{E}\left\{ \frac{c(i)-\xi_n}{\left[A_n + \frac{\gamma_n\xi_n}{N-1}\right]\left[A_n + \frac{\gamma_n c_i}{N-1}\right]} /\mathcal{F}_{n-1} \bigwedge u_n = u(i) \right\} p_n(i) \overset{a.s.}{=}$$

$$= \gamma_{nN} \sum_{i\neq\alpha}^{N} \mathbf{E}\left\{ \frac{B_n\left[1 - \frac{\gamma_n\xi_n}{A_n(N-1)} + o(\frac{\gamma_n}{p_n(\alpha)})\right]}{A_n\left[A_n + \frac{\gamma_n c_i}{N-1}\right]} /\mathcal{F}_{n-1} \bigwedge u_n = u(i) \right\} p_n(i) \overset{a.s.}{=}$$

$$= \gamma_{nN} \sum_{i\neq\alpha}^{N} \frac{\mathbf{E}\left\{ B_n\left[-\frac{\gamma_n\xi_n}{A_n(N-1)} + o(\frac{\gamma_n}{p_n(\alpha)})\right] /\mathcal{F}_{n-1} \bigwedge u_n = u(i) \right\}}{A_n\left[A_n + \frac{\gamma_n c_i}{N-1}\right]} p_n(i)$$

where

$$\gamma_{nN} = \frac{\gamma_n}{N-1}$$

$$A_n = p_n(\alpha)(1-\gamma_n)$$

and

$$B_n = (c(i) - \xi_n)$$

Using then Cauchy-Bounyakovskii inequality we get

$$s_n \leq \frac{\gamma_n^2 C}{(N-1)^2} \sum_{i\neq\alpha}^{N} \frac{1}{p_n^2(\alpha)(1-\gamma_n)^2\left[p_n(\alpha)(1-\gamma_n) + \frac{\gamma_n c_i}{N-1}\right]} p_n(i) \leq$$

$$\leq \frac{\gamma_n^2 C}{(N-1)^2} \frac{1}{p_n^2(\alpha)(1-\gamma_n)^2\left[p_n(\alpha)(1-\gamma_n) + \frac{\gamma_n c^-}{N-1}\right]} \sum_{i\neq\alpha}^{N} p_n(i) = \tag{3.27}$$

$$= \frac{\gamma_n^2 C}{(N-1)^2} \frac{W_n}{p_n(\alpha)(1-\gamma_n)^2\left[p_n(\alpha)(1-\gamma_n) + \frac{\gamma_n c^-}{N-1}\right]}$$

where

$$C := \max_{i\neq\alpha} \sqrt{\mathbf{E}\left\{ (c(i)-\xi_n)^2 /\mathcal{F}_{n-1} \bigwedge u_n = u(i) \right\}} = \max_{i\neq\alpha} \sigma_i$$

$$\sigma_i^2 := \mathbf{E}\left\{ (c(i)-\xi_n)^2 /\mathcal{F}_{n-1} \bigwedge u_n = u(i) \right\}$$

$$c^- := \max_{i\neq\alpha} c(i)$$

Using the lower estimations (3.22) and (3.23) in (3.27), we finally obtain

$$s_n \leq \frac{\gamma_n^2 C}{(N-1)^2} \frac{W_n}{p_n(\alpha)(1-\gamma_n)^2 \left[p_n(\alpha)(1-\gamma_n) + \frac{\gamma_n c^-}{N-1}\right]} \leq \tag{3.28}$$

$$\leq \frac{\gamma_n^2 C}{(N-1)^2} \frac{W_n}{p_n(\alpha) \min\left\{p_n(\alpha); \frac{c^-}{N-1}\right\}} \leq$$

$$\leq \frac{\gamma^2 C}{(N-1)^2} \frac{W_n}{(a-\gamma)^{2\gamma}} \frac{1}{(n+a)^{2(1-\gamma)}} = \tag{3.29}$$

$$= n^{-2(1-\gamma)} C_1 (1+o(1)) W_n$$

$$C_1 := \frac{\gamma^2 C}{(a-\gamma)^{2\gamma} (N-1)^2}$$

for $n \geq n_0(\omega)$, $n_0(\omega) \overset{a.s.}{<} \infty$.
Let us estimate the first term in (3.24):

$$\sum_{i=1}^{N} \left(\frac{1}{\mathbf{E}\{p_{n+1}(\alpha)/\mathcal{F}_{n-1} \bigwedge u_n = u(i)\}} - 1 \right) p_n(i) =$$

$$= \sum_{i \neq \alpha}^{N} \left(\frac{1}{\mathbf{E}\{p_{n+1}(\alpha)/\mathcal{F}_{n-1} \bigwedge u_n = u(i)\}} - 1 \right) p_n(i) +$$

$$+ \left(\frac{1}{\mathbf{E}\{p_{n+1}(\alpha)/\mathcal{F}_{n-1} \bigwedge u_n = u(\alpha)\}} - 1 \right) p_n(\alpha) =$$

$$= \sum_{i \neq \alpha}^{N} \left(\frac{1}{\left[p_n(\alpha)(1-\gamma_n) + \gamma_n \frac{c^-}{N-1}\right]} - 1 \right) p_n(i) +$$

$$+ \left(\frac{1}{[p_n(\alpha)(1-\gamma_n) + \gamma_n]} - 1 \right) p_n(\alpha) \leq$$

$$\leq \left(\frac{1}{\left[p_n(\alpha)(1-\gamma_n) + \gamma_n \frac{c^-}{N-1}\right]} - 1 \right) (1 - p_n(\alpha)) +$$

$$+ \frac{(1-p_n(\alpha))(1-\gamma_n)}{p_n(\alpha)(1-\gamma_n) + \gamma_n} p_n(\alpha) \leq$$

$$\leq \left(\frac{1}{\left[p_n(\alpha)(1-\gamma_n) + \gamma_n \frac{c^-}{N-1}\right]} - 1 \right) (1 - p_n(\alpha)) +$$

$$+\frac{(1-p_n(\alpha))(1-\gamma_n)}{p_n(\alpha)(1-\gamma_n)+\gamma_n\frac{c^-}{N-1}}p_n(\alpha)=$$

$$=\frac{p_n(\alpha)}{p_n(\alpha)(1-\gamma_n)+\gamma_n\frac{c^-}{N-1}}\left[1-\gamma_n\frac{c^-}{N-1}\right]W_n=$$

$$=\frac{1}{(1-\gamma_n)+\frac{\gamma_n}{p_n(\alpha)}\frac{c^-}{N-1}}\left[1-\gamma_n\frac{c^-}{N-1}\right]W_n$$

Using the estimations (3.22) and (3.23), we derive:

$$\sum_{i=1}^{N}\left(\frac{1}{\mathbf{E}\{p_{n+1}(\alpha)/\mathcal{F}_{n-1}\bigwedge u_n=u(i)\}}-1\right)p_n(i)\le$$

$$\le\frac{1}{(1-\gamma_n)+\frac{\gamma(a-\gamma)^{-\gamma}}{(n+a)^{1-\gamma}p_1(\alpha)}\frac{c^-}{N-1}}\left[1-\gamma_n\frac{c^-}{N-1}\right]W_n=$$

$$=\frac{1}{1+n^{-(1-\gamma)}C_2(1+o(1))}\left[1-\gamma_n\frac{c^-}{N-1}\right]W_n=$$

$$=\left[1-n^{-(1-\gamma)}C_2(1+o(1))\right]\left[1-\gamma_n\frac{c^-}{N-1}\right]W_n= \tag{3.30}$$

$$=\left[1-n^{-(1-\gamma)}C_2(1+o(1)+O(n^{-\gamma}))\right]W_n=$$

$$=\left[1-n^{-(1-\gamma)}C_2(1+o(1))\right]W_n$$

with

$$C_2=\frac{\gamma c^-}{[N-1]\,p_1(\alpha)\,(a-\gamma)^{\gamma}}$$

Combining (3.24), (3.29) and (3.30), we finally obtain:

$$\mathbf{E}\{W_{n+1}/\mathcal{F}_n\}\overset{a.s.}{\le}$$

$$\overset{a.s.}{\le}W_n\left[1-n^{-(1-\gamma)}C_2(1+o(1))+n^{-2(1-\gamma)}C_1\,(1+o(1))\right]=$$

$$=W_n\left[1-n^{-(1-\gamma)}C_2(1+o(1))\right] \tag{3.31}$$

with

$$C_1:=\frac{\gamma^2\max\limits_{i\ne\alpha}\sigma_i}{(a-\gamma)^{2\gamma}\,(N-1)^2},\quad C_2=\frac{N\gamma c^-}{[N-1]\,(a-\gamma)^{\gamma}} \tag{3.32}$$

Appealing to (3.31) and Robbins-Siegmund theorem [23] we obtain the convergence with probability one. Theorem is proved. ■

Corollary 2. *Under the assumptions of this theorem the convergence rate is described by*

$$\mathbf{E}\{W_{n+1}\} \leq (N-1)\prod_{t=1}^{n}\left[1-t^{-(1-\gamma)}C_2(1+o(1))\right] < \tag{3.33}$$

$$< (N-1)\exp\{-(C_2-\varepsilon)\,n^{\gamma}\}$$

$$0 < \varepsilon \text{ small enough}$$

Proof.

This result follows directly from (3.31) after averaging and using the following inequalities

$$\prod_{t=1}^{n}\left[1-t^{-(1-\gamma)}C_2(1+o(1))\right] \leq \prod_{t=1}^{n}\left[1-t^{-(1-\gamma)}\,(C_2-\varepsilon)\right]$$

and

$$\prod_{t=1}^{n}\left[1-t^{-(1-\gamma)}C\right] \leq \exp\{-(C-\varepsilon)\,n^{\gamma}\}$$

Corollary is proved. ■

The message in this theorem is that a learning automaton using the Bush-Mosteller reinforcement scheme selects asymptotically the optimal action $u_n = u(\alpha)$, i.e., its behaviour is asymptotically optimal. The learning process rate increases when the parameter γ in the reinforcement scheme (3.21) (see assumption 2 of this theorem) is selected close to 1.

The next subsection deals with the analysis of Shapiro-Narendra reinforcement scheme.

3.4.2 Shapiro-Narendra reinforcement scheme

Consider now the following reinforcement scheme [19]-[14]:

$$p_{n+1} = p_n + \gamma_n(1-\xi_n)\left[e(u_n)-p_n\right] \tag{3.34}$$

with

$$\begin{aligned}\gamma_n &\in [0,1]\,, \quad \xi_n \in [0,1]\,, p_1(i) = 1/N \\ e(u_n) &= (\underbrace{0,...,0,1}_{i},0,...,0)^T\,,\ u_n = u(i) \\ e^N &= (1,...,1)^T \in R^N\end{aligned}$$

We are now ready for our main results.

Theorem 4. *If*

1. *the environment response sequence* $\{\xi_n\}$ *satisfies assumptions (H1),(H2)*
2. *the optimal action* $u_n = u(\alpha)$ *is unique, i.e.,*

$$c^- := \min_{i \neq \alpha} c(i) > c(\alpha) \geq 0$$

3. *the correction factor is selected as follows:*

$$\gamma_n = \frac{\gamma}{n+a}, \ \gamma \in (0,1), a > \gamma$$

Then, the Lyapunov function (3.20) possesses the following property

$$W(p_n) \underset{n \to \infty}{\overset{a.s.}{\to}} 0$$

and, hence, the reinforcement scheme (3.34) generates asymptotically an optimal sequence $\{p_n\}$*, such that*

$$p_n \underset{n \to \infty}{\overset{a.s.}{\to}} p^*$$

Proof.
We follow the lines of the proof described above. Let us consider the following estimation:

$$p_{n+1}(i) = p_n(i) + \gamma_n (1 - \xi_n) [\chi(u_n = u(i)) - p_n(i)] =$$

$$= p_n(i) [1 - \gamma_n (1 - \xi_n)] + \gamma_n (1 - \xi_n) \chi(u_n = u(i)) \geq \tag{3.35}$$

$$\geq p_n(i) (1 - \gamma_n) \geq \cdots \geq p_1(i) \prod_{t=1}^{n} (1 - \gamma_t) > 0$$

The reinforcement scheme (3.34) leads to

$$\mathbf{E}\left\{\frac{1 - p_{n+1}(\alpha)}{p_{n+1}(\alpha)} / \mathcal{F}_n\right\} \overset{a.s.}{=}$$

$$\overset{a.s.}{=} \sum_{i=1}^{N} \mathbf{E}\left\{\frac{1}{p_{n+1}(\alpha)} - 1 / \mathcal{F}_{n-1} \bigwedge u_n = u(i)\right\} p_n(i) \overset{a.s.}{=}$$

$$= \sum_{i=1}^{N} \left(\frac{1}{\mathbf{E}\{p_{n+1}(\alpha) / \mathcal{F}_{n-1} \bigwedge u_n = u(i)\}} - 1 \right) p_n(i) + s_n \tag{3.36}$$

where

$$s_n = \sum_{i=1}^{N} \mathbf{E}\left\{ \frac{1}{p_{n+1}(\alpha)} - \right.$$

$$\left. - \frac{1}{\mathbf{E}\left\{p_{n+1}(\alpha)/\mathcal{F}_{n-1} \bigwedge u_n = u(i)\right\}} / \mathcal{F}_{n-1} \bigwedge u_n = u(i) \right\} p_n(i) \qquad (3.37)$$

Let us now estimate the term s_n (3.37):

$$s_n = \sum_{i=1}^{N} \mathbf{E}\left\{ \frac{E_\alpha - p_{n+1}(\alpha)}{p_{n+1}(\alpha) E_\alpha} / \mathcal{F}_{n-1} \bigwedge u_n = u(i) \right\} p_n(i) \overset{a.s.}{=}$$

$$= \gamma_n \sum_{i=1}^{N} \mathbf{E}\left\{ \frac{C_i}{p_n(\alpha) + \gamma_n(1-\xi_n)\left(\delta_{i,\alpha} - p_n(\alpha)\right)} \times \right.$$

$$\left. \times \frac{\left(\delta_{i,\alpha} - p_n(\alpha)\right)}{p_n(\alpha) + \gamma_n(1-c(i))\left(\delta_{i,\alpha} - p_n(\alpha)\right)} / \mathcal{F}_{n-1} \bigwedge u_n = u(i) \right\} p_n(i) \overset{a.s.}{=}$$

$$= \gamma_n \sum_{i=1}^{N} \mathbf{E}\left\{ \frac{C_i \left[1 - \frac{\gamma_n \xi_n (\delta_{i,\alpha} - p_n(\alpha))}{p_n(\alpha) + \gamma_n(\delta_{i,\alpha} - p_n(\alpha))} + o\left(\frac{\gamma_n \xi_n (\delta_{i,\alpha} - p_n(\alpha))}{p_n(\alpha) + \gamma_n(\delta_{i,\alpha} - p_n(\alpha))}\right)\right]}{p_n(\alpha) + \gamma_n\left(\delta_{i,\alpha} - p_n(\alpha)\right)} \times \right.$$

$$\left. \times \frac{\left(\delta_{i,\alpha} - p_n(\alpha)\right)}{p_n(\alpha) + \gamma_n(1-c(i))\left(\delta_{i,\alpha} - p_n(\alpha)\right)} / \mathcal{F}_{n-1} \bigwedge u_n = u(i) \right\} p_n(i) \overset{a.s.}{=}$$

$$= \gamma_n \sum_{i \neq \alpha}^{N} \mathbf{E}\left\{ \frac{C_i \left[\frac{\gamma_n \xi_n}{(1-\gamma_n)} + o\left(\frac{\gamma_n}{(1-\gamma_n)}\right)\right]}{p_n(\alpha)(1-\gamma_n)} \times \right.$$

$$\left. \times \frac{(-1)}{\left[1 - \gamma_n(1-c(i))\right]} / \mathcal{F}_{n-1} \bigwedge u_n = u(i) \right\} p_n(i) +$$

$$+ \gamma_n \mathbf{E}\left\{ \frac{C_\alpha \left[-\frac{\gamma_n \xi_n (1 - p_n(\alpha))}{p_n(\alpha) + \gamma_n(1 - p_n(\alpha))} + o\left(\frac{\gamma_n \xi_n (1 - p_n(\alpha))}{p_n(\alpha) + \gamma_n(1 - p_n(\alpha))}\right)\right]}{p_n(\alpha) + \gamma_n\left(1 - p_n(\alpha)\right)} \times \right.$$

$$\left. \times \frac{\left(1 - p_n(\alpha)\right)}{p_n(\alpha) + \gamma_n(1-c(\alpha))\left(1 - p_n(\alpha)\right)} / \mathcal{F}_{n-1} \bigwedge u_n = u(\alpha) \right\} p_n(\alpha)$$

where

$$E_\alpha = \mathbf{E}\left\{p_{n+1}(\alpha)/\mathcal{F}_{n-1} \bigwedge u_n = u(i)\right\}$$

$$C_i = (c(i) - \xi_n)$$

and

$$C_\alpha = (c(\alpha) - \xi_n)$$

Using the Cauchy-Bounyakovskii inequality, it follows

$$s_n \overset{a.s.}{\leq} \frac{\gamma_n^2}{p_n(\alpha)(1-\gamma_n)^2[1-\gamma_n(1-c^-)]}\times$$

$$\times \sum_{i\neq\alpha}^{N} \mathbf{E}\left\{|c(i)-\xi_n|[1+o(1)]/\mathcal{F}_{n-1}\bigwedge u_n=u(i)\right\}p_n(i)+$$

$$+\gamma_n^2 \frac{p_n(\alpha)(1-p_n(\alpha))^2}{[p_n(\alpha)+\gamma_n(1-p_n(\alpha))]^2[p_n(\alpha)+\gamma_n(1-c(\alpha))(1-p_n(\alpha))]}\times$$

$$\times\mathbf{E}\left\{|c(\alpha)-\xi_n|[1+o(1)]/\mathcal{F}_{n-1}\bigwedge u_n=u(\alpha)\right\}\leq$$

$$\leq \gamma_n^2 C[1+o(1)]W_n\left[\frac{1}{(1-\gamma_n)^2[1-\gamma_n(1-c^-)]}+\right.$$

$$\left.+\frac{p_n^3(\alpha)W_n}{[p_n(\alpha)+\gamma_n(1-p_n(\alpha))]^2[p_n(\alpha)+\gamma_n(1-c(\alpha))(1-p_n(\alpha))]}\right]\leq$$

$$\leq \gamma_n^2 C[1+o(1)]W_n(1+W_n)=\frac{\gamma_n^2}{p_n(\alpha)}C[1+o(1)]W_n \tag{3.38}$$

where C, σ_i^2 and c^- are defined as follows:

$$C:=\max_{i\neq\alpha}\sqrt{\mathbf{E}\left\{(c(i)-\xi_n)^2/\mathcal{F}_{n-1}\bigwedge u_n=u(i)\right\}}=\max_{i\neq\alpha}\sigma_i$$

$$\sigma_i^2:=\mathbf{E}\left\{(c(i)-\xi_n)^2/\mathcal{F}_{n-1}\bigwedge u_n=u(i)\right\}$$

$$c^-:=\max_{i\neq\alpha}c(i)$$

Using inequalities (3.23) and (3.34) we derive:

$$\gamma_n p_n^{-1}(\alpha)\leq p_1^{-1}(\alpha)\frac{\gamma}{n+a}\prod_{t=1}^{n}(1-\frac{\gamma}{t+a})^{-1}\leq \tag{3.39}$$

$$\leq p_1^{-1}(\alpha)\frac{\gamma}{n+a}\frac{(n+a)^\gamma}{(a-\gamma)^\gamma}=\frac{p_1^{-1}(\alpha)\gamma[1+o(1)]}{(a-\gamma)^\gamma}n^{-1+\gamma}$$

We finally obtain:

$$s_n \overset{a.s.}{\leq} n^{-2+\gamma}C_1W_n \tag{3.40}$$

with

$$C_1:=\frac{p_1^{-1}(\alpha)\gamma^2[1+o(1)]}{(a-\gamma)^\gamma}C \tag{3.41}$$

Let us estimate the first term in (3.36):

$$\sum_{i=1}^{N}\left(\frac{1}{\mathbf{E}\left\{p_{n+1}(\alpha)/\mathcal{F}_{n-1}\bigwedge u_n=u(i)\right\}}-1\right)p_n(i)=$$

$$=\sum_{i=1}^{N}\left(\frac{1}{p_n(\alpha)+\gamma_n(1-c(i))\left(\delta_{i,\alpha}-p_n(\alpha)\right)}-1\right)p_n(i)=$$

$$=\sum_{i\neq\alpha}^{N}\left(\frac{1}{[p_n(\alpha)(1-\gamma_n(1-c(i)))]}-1\right)p_n(i)+$$

$$+\left(\frac{1}{[p_n(\alpha)+\gamma_n(1-c(\alpha))\left(1-p_n(\alpha)\right)]}-1\right)p_n(\alpha)\leq$$

$$\leq\sum_{i\neq\alpha}^{N}\left(\frac{1}{p_n(\alpha)(1-\gamma_n(1-c^-))}-1\right)p_n(i)+$$

$$+\frac{p_n(\alpha)\left(1-p_n(\alpha)\right)\left[1-\gamma_n(1-c(\alpha))\right]}{p_n(\alpha)+\gamma_n(1-c(\alpha))\left(1-p_n(\alpha)\right)}$$

Using the Lyapunov function (3.20) we obtain:

$$\sum_{i=1}^{N}\left(\frac{1}{\mathbf{E}\left\{p_{n+1}(\alpha)/\mathcal{F}_{n-1}\bigwedge u_n=u(i)\right\}}-1\right)p_n(i)\leq$$

$$\leq W_n\left[\frac{1-p_n(\alpha)(1-\gamma_n(1-c^-))}{1-\gamma_n(1-c^-)}+\frac{p_n^2(\alpha)\left[1-\gamma_n(1-c(\alpha))\right]}{p_n(\alpha)+\gamma_n(1-c(\alpha))\left(1-p_n(\alpha)\right)}\right] \tag{3.42}$$

These terms can be expanded via Taylor series about γ_n, i.e.,

$$\frac{1-p_n(\alpha)(1-\gamma_n(1-c^-))}{1-\gamma_n(1-c^-)}=1+\gamma_n(1-c^-)-$$

$$-p_n(\alpha)\left[1-\gamma_n^2(1-c^-)^2\right]+O(\gamma_n^2)$$

$$\frac{p_n^2(\alpha)\left[1-\gamma_n(1-c(\alpha))\right]}{p_n(\alpha)+\gamma_n(1-c(\alpha))\left(1-p_n(\alpha)\right)}=p_n(\alpha)-\gamma_n(1-c(\alpha))+O(\frac{\gamma_n^2}{p_n(\alpha)})$$

Substituting these relations into (3.42) we obtain:

$$\sum_{i=1}^{N}\left(\frac{1}{\mathbf{E}\left\{p_{n+1}(\alpha)/\mathcal{F}_{n-1}\bigwedge u_n=u(i)\right\}}-1\right)p_n(i)\leq$$

$$\leq W_n\left(1-\gamma_n\left[c^- - c(\alpha)\right]+O(\frac{\gamma_n^2}{p_n(\alpha)})\right)$$

If the estimations (3.22) and (3.23) are used, the following is obtained:

$$\sum_{i=1}^{N}\left(\frac{1}{\mathbf{E}\{p_{n+1}(\alpha)/\mathcal{F}_{n-1}\bigwedge u_n = u(i)\}}-1\right)p_n(i)\leq$$

$$\leq W_n\left(1-\frac{\gamma}{n+a}\left[c^- - c(\alpha)\right]+O(\frac{\gamma_n^2}{p_1(\alpha)\left(\frac{a-\gamma}{n+a}\right)^{\gamma}})\right)=$$

$$=W_n\left(1-\frac{\gamma}{n+a}\left[c^- - c(\alpha)\right]+O(n^{-2+\gamma})\right) \tag{3.43}$$

Combining (3.36), (3.40) and (3.43), we finally obtain:

$$\mathbf{E}\{W_{n+1}/\mathcal{F}_n\}\overset{a.s.}{\leq}W_n\left(1-\frac{\gamma}{n+a}\left[c^- - c(\alpha)\right]+O(n^{-2+\gamma})+n^{-2+\gamma}C_1\right)=$$

$$=W_n\left(1-\frac{\gamma}{n+a}\left[c^- - c(\alpha)\right]+O(n^{-2+\gamma})\right) \tag{3.44}$$

From (3.44) and Robbins-Siegmund theorem [23] we obtain the convergence with probability one. Theorem is proved. ■

Corollary 3. *Under the assumptions of this theorem the convergence rate is described by the formula*

$$\mathbf{E}\{W_{n+1}\}\leq \tag{3.45}$$

$$\leq (N-1)\prod_{t=1}^{n}\left(1-\frac{\gamma}{t+a}\left[c^- - c(\alpha)\right](1+o(1))\right)<$$

$$<(N-1)\,n^{-C}$$

$$C=\left[\gamma-\varepsilon\right]\left[c^- - c(\alpha)\right],\ 0<\varepsilon \text{ small enough}$$

Proof.

This result follows directly from (3.44) after averaging and using the following inequalities

$$\prod_{t=1}^{n}\left(1-\frac{\gamma}{t+a}\left[c^- - c(\alpha)\right](1+o(1))\right)\leq\prod_{t=1}^{n}\left(1-\frac{1}{t}\left[\gamma-\varepsilon\right]\left[c^- - c(\alpha)\right]\right)$$

and

$$\prod_{t=1}^{n}\left[1-t^{-1}C\right]\leq\exp\left\{-C\sum_{t=1}^{n}t^{-1}\right\}\leq\exp\{-C\ln n\}=n^{-C}$$

Corollary is proved. ■

From this theorem and its corollary it follows that a learning automaton using the Shapiro-Narendra reinforcement scheme and operating in random stationary environments with nonbinary responses, belonging to the unit segment, selects asymptotically the optimal action $u_n = u(\alpha)$, i.e., its behaviour is asymptotically optimal. The learning process rate increases when the parameter γ in the reinforcement scheme (3.34) (see assumption 2 of this theorem) is selected close to 1. It also increases when the difference $[c^- - c(\alpha)]$ increases.

The next subsection deals with the analysis of the Varshavskii-Vorontsova reinforcement scheme [20].

3.4.3 Varshavskii-Vorontsova reinforcement scheme

Consider now the reinforcement scheme, described in [20]-[14]:

$$p_{n+1} = p_n + \gamma_n p_n^T \, e(u_n)(1 - 2\xi_n) \left[e(u_n) - p_n \right] \tag{3.46}$$

with

$$\begin{aligned} \gamma_n &\in [0,1], \quad \xi_n \in [0,1], p_1(i) = 1/N \\ e(u_n) &= (\underbrace{0,...,0,1}_{i},0,...,0)^T, \; u_n = u(i) \\ e^N &= (1,...,1)^T \in R^N \end{aligned}$$

In comparison with the two previous reinforcement schemes, which are linear with respect to the vector p_n, this scheme is nonlinear (quadratic) and as a consequence its properties differ from those stated in the previous subsections. In spite of its nonlinearity characteristic, the convergence analysis of this scheme can be done on the basis of the Lyapunov approach using the Lyapunov function (3.20).

Theorem 5. *If*

1. *the environment response sequence* $\{\xi_n\}$ *satisfies assumptions (H1),(H2)*
2. *the optimal action* $u_n = u(\alpha)$ *is unique and all the nonoptimal average losses (corresponding to the non optimal actions) are great than* 1/2, *i.e.,* $c^- := \min_{i \neq \alpha} c(i) > \frac{1}{2} > c(\alpha) \geq 0$
3. *the correction factor (learning rate) is selected as follows:*

$$\gamma_n \in \left(0, \min\left\{ \left[\min\{|b|;(1-2c(\alpha))\}\right]^{-1}; 1-\varepsilon \right\}\right), \; \sum_{n=1}^{\infty} \gamma_n = \infty$$

$$0 < \varepsilon \text{ - } small\ enough$$

then, the Lyapunov function (3.20) possesses the property

$$W(p_n) \underset{n \to \infty}{\overset{a.s.}{\to}} 0$$

and, hence, the reinforcement scheme (3.46) generates asymptotically an optimal sequence $\{p_n\}$, such that

$$p_n \underset{n \to \infty}{\overset{a.s.}{\to}} p^*$$

This theorem is proven in the same manner as the previous one.

Proof.

The reinforcement scheme (3.46) leads to

$$\mathbf{E}\left\{\frac{1 - p_{n+1}(\alpha)}{p_{n+1}(\alpha)}/\mathcal{F}_n\right\} \overset{a.s.}{=} \sum_{i=1}^{N} \mathbf{E}\left\{\frac{1}{p_{n+1}(\alpha)} - 1/\mathcal{F}_{n-1} \bigwedge u_n = u(i)\right\} \times$$

$$\times p_n(i) \overset{a.s.}{=}$$

$$= \sum_{i=1}^{N} \left(\frac{1}{\mathbf{E}\left\{p_{n+1}(\alpha)/\mathcal{F}_{n-1} \bigwedge u_n = u(i)\right\}} - 1 \right) p_n(i) + s_n \tag{3.47}$$

where

$$s_n = \sum_{i=1}^{N} \mathbf{E}\left\{\frac{1}{p_{n+1}(\alpha)} - \right.$$

$$\left. - \frac{1}{\mathbf{E}\left\{p_{n+1}(\alpha)/\mathcal{F}_{n-1} \bigwedge u_n = u(i)\right\}}/\mathcal{F}_{n-1} \bigwedge u_n = u(i)\right\} p_n(i) \tag{3.48}$$

Let us now estimate the term s_n (3.48):

$$s_n = \sum_{i=1}^{N} \mathbf{E}\left\{\frac{E_\alpha - p_{n+1}(\alpha)}{p_{n+1}(\alpha) E_\alpha}/\mathcal{F}_{n-1} \bigwedge u_n = u(i)\right\} p_n(i) \overset{a.s.}{=}$$

$$= 2\gamma_n \sum_{i=1}^{N} \mathbf{E}\left\{\frac{p_n(i)\left(\xi_n - c(i)\right)}{p_n(\alpha) + \gamma_n(1 - 2\xi_n)\left(\delta_{i,\alpha} - p_n(\alpha)\right) p_n(i)} \times \right.$$

$$\left. \times \frac{\left(\delta_{i,\alpha} - p_n(\alpha)\right)}{p_n(\alpha) + \gamma_n(1 - 2c(i))\left(\delta_{i,\alpha} - p_n(\alpha)\right) p_n(i)}/\mathcal{F}_{n-1} \bigwedge u_n = u(i)\right\} p_n(i) \overset{a.s.}{=}$$

$$= 2\gamma_n \sum_{i=1}^{N} \mathbf{E}\left\{\frac{p_n(i)\left(\xi_n - c(i)\right)\left[1 - C_n + o(C_n)\right]}{p_n(\alpha) + \gamma_n\left(\delta_{i,\alpha} - p_n(\alpha)\right) p_n(i)} \times \right.$$

$$\times \frac{(\delta_{i,\alpha} - p_n(\alpha))}{p_n(\alpha) + \gamma_n(1-2c(i))\,(\delta_{i,\alpha} - p_n(\alpha))\,p_n(i)} / \mathcal{F}_{n-1} \bigwedge u_n = u(i) \Big\} p_n(i) \overset{a.s.}{=}$$

$$= 4\gamma_n \sum_{i \neq \alpha}^{N} \mathbf{E} \left\{ \frac{p_n(i)\,(c(i) - \xi_n) \left[\frac{\gamma_n \xi_n p_n(i)}{[1-\gamma_n p_n(i)]} + o(\frac{\gamma_n \xi_n p_n(i)}{[1-\gamma_n p_n(i)]}) \right]}{p_n(\alpha)\,(1 - \gamma_n p_n(i))} \times \right.$$

$$\left. \times \frac{1}{1 - \gamma_n\,(1-2c(i))\,p_n(i)} / \mathcal{F}_{n-1} \bigwedge u_n = u(i) \right\} p_n(i) +$$

$$+4\gamma_n \mathbf{E} \left\{ \frac{(\xi_n - c(\alpha)) \left[-\frac{\gamma_n \xi_n (1-p_n(\alpha))}{1+\gamma_n(1-p_n(\alpha))} + o(\frac{\gamma_n \xi_n (1-p_n(\alpha))}{1+\gamma_n(1-p_n(\alpha))}) \right]}{1 + \gamma_n\,(1 - p_n(\alpha))} \times \right.$$

$$\left. \times \frac{(1 - p_n(\alpha))}{1 + \gamma_n(1-2c(\alpha))\,(1-p_n(\alpha))} / \mathcal{F}_{n-1} \bigwedge u_n = u(\alpha) \right\} \overset{a.s.}{\leq}$$

$$\leq 4\gamma_n^2 \frac{[1+o(1)]}{p_n(\alpha)} \sum_{i \neq \alpha}^{N} \frac{p_n^2(i)}{(1 - \gamma_n p_n(i))} \frac{\mathbf{E}\left\{ |c(i) - \xi_n| / \mathcal{F}_{n-1} \bigwedge u_n = u(i) \right\}}{[1 - \gamma_n\,(1-2c(i))\,p_n(i)]} +$$

$$+4\gamma_n^2 \frac{[1+o(1)]\,(1-p_n(\alpha))\,\mathbf{E}\left\{ |\xi_n - c(\alpha)| / \mathcal{F}_{n-1} \bigwedge u_n = u(\alpha) \right\}}{[1+\gamma_n(1-2c(\alpha))\,(1-p_n(\alpha))]\,[1+\gamma_n\,(1-p_n(\alpha))]} \tag{3.49}$$

where

$$E_\alpha = \mathbf{E}\left\{ p_{n+1}(\alpha) / \mathcal{F}_{n-1} \bigwedge u_n = u(i) \right\}$$

and

$$C_n = \frac{2\gamma_n \xi_n\,(\delta_{i,\alpha} - p_n(\alpha))\,p_n(i)}{p_n(\alpha) + \gamma_n\,(\delta_{i,\alpha} - p_n(\alpha))\,p_n(i)}$$

Using the Cauchy-Bounyakovskii inequality, from (3.49) we get the following upper estimation:

$$s_n \overset{a.s.}{\leq} 4\gamma_n^2 C\,[1+o(1)]\,W_n \left[\frac{1}{(1-\gamma_n)} + p_n(\alpha) \right] \leq$$

$$\leq 8\gamma_n^2 C\,[1+o(1)]\,W_n \tag{3.50}$$

where

$$C := \max_{i \neq \alpha} \sqrt{\mathbf{E}\left\{ (c(i) - \xi_n)^2 / \mathcal{F}_{n-1} \bigwedge u_n = u(i) \right\}} = \max_{i \neq \alpha} \sigma_i$$

$$\sigma_i^2 := \mathbf{E}\left\{ (c(i) - \xi_n)^2 / \mathcal{F}_{n-1} \bigwedge u_n = u(i) \right\}$$

$$c^- := \max_{i \neq \alpha} c(i)$$

Let us now calculate the first term in (3.47):

$$\sum_{i=1}^{N} \left(\frac{1}{\mathbf{E}\left\{ p_{n+1}(\alpha) / \mathcal{F}_{n-1} \bigwedge u_n = u(i) \right\}} - 1 \right) p_n(i) \overset{a.s.}{=}$$

$$= \sum_{i=1}^{N} \frac{p_n(i)}{p_n(\alpha) + \gamma_n(1 - c(i))\,(\delta_{i,\alpha} - p_n(\alpha))\,p_n(i)} - 1 =$$

$$= \frac{1}{p_n(\alpha)} \sum_{i \neq \alpha}^{N} \frac{p_n(i)}{[1 - \gamma_n(1 - 2c(i))p_n(i)]} + \tag{3.51}$$

$$+ \frac{1}{1 + \gamma_n(1 - 2c(\alpha))\,(1 - p_n(\alpha))} - 1$$

Let us now maximize the first term in (3.51) with respect to the components $p_n(i)$ $(i \neq \alpha)$ under the constraint

$$\sum_{i \neq \alpha}^{N} p_n(i) = 1 - p_n(\alpha) \tag{3.52}$$

To do that, let us introduce the following variables

$$x_i := p_n(i), \quad a_i := \gamma_n\,[1 - 2c(i)]\,, \quad i \neq \alpha$$

It follows that

$$\sum_{i \neq \alpha}^{N} \frac{p_n(i)}{1 - \gamma_n\,[1 - 2c(i)]\;p_n(i)} = \sum_{i \neq \alpha}^{N} \frac{x_i}{1 - a_i x_i} := F(x) \tag{3.53}$$

To maximize this function (3.53) under the constraint (3.52), let us introduce the following Lagrange function:

$$L(x, \lambda) := F(x) - \lambda \left[\sum_{i \neq \alpha}^{N} x_i - (1 - x_\alpha) \right]$$

The optimal solution (x^*, λ^*) satisfies the following conditions of optimality:

$$\frac{\partial}{\partial x_i} L(x^*, \lambda^*) = \frac{1}{(1 - a_i x_i^*\,)^2} - \lambda^* = 0 \quad \forall i \neq \alpha$$

$$\frac{\partial}{\partial \lambda} L(x^*, \lambda^*) = \sum_{i \neq \alpha}^{N} x_i^* - (1 - x_\alpha^*) = 0$$

From these optimality conditions, it follows that

$$x_i^* = \frac{1}{a_i} \frac{1 - x_\alpha}{\sum_{i \neq \alpha}^{N} a_i^{-1}}, \qquad \sqrt{\lambda^*} = (1 - \frac{1 - x_\alpha}{\sum_{i \neq \alpha}^{N} a_i^{-1}})^{-1}$$

Hence

$$F(x) \leq F(x^*) = (1-x_\alpha)\,(1-\frac{1-x_\alpha}{\sum\limits_{i\neq\alpha}^{N} a_i^{-1}})^{-1}$$

From this inequality we derive

$$\sum_{i\neq\alpha}^{N}\left(\frac{p_n(i)}{1-\gamma_n\,[1-2c(i)]\;p_n(i)}\right) \leq \frac{1-\;p_n(\alpha)}{1-\gamma_n b\,[1-\;p_n(\alpha)]} \tag{3.54}$$

where

$$b := \left(\sum_{i\neq\alpha}^{N}[1-2c(i)]^{-1}\right)^{-1} \tag{3.55}$$

Notice that according to the first assumption of this theorem

$$b < 0$$

So, we can rewrite (3.54) as follows

$$\sum_{i\neq\alpha}^{N}\left(\frac{p_n(i)}{1-\gamma_n\,[1-2c(i)]\;p_n(i)}\right) \leq \frac{1-\;p_n(\alpha)}{1+\gamma_n\,|b|\,[1-\;p_n(\alpha)]} \tag{3.56}$$

Substituting (3.56) into (3.51) leads to:

$$\sum_{i=1}^{N}\left(\frac{1}{\mathbf{E}\left\{p_{n+1}(\alpha)/\mathcal{F}_{n-1}\bigwedge u_n = u(i)\right\}}-1\right)p_n(i) \overset{a.s.}{\leq}$$

$$\leq \frac{1}{p_n(\alpha)}\frac{1-\;p_n(\alpha)}{1+\gamma_n\,|b|\,[1-\;p_n(\alpha)]}+$$

$$-\frac{\gamma_n(1-2c(\alpha))\,(1-p_n(\alpha))}{1+\gamma_n(1-2c(\alpha))\,(1-p_n(\alpha))}=$$

$$= W_n\left[\frac{1}{1+\gamma_n\,|b|\,[1-\;p_n(\alpha)]}-\frac{\gamma_n(1-2c(\alpha))p_n(\alpha)}{1+\gamma_n(1-2c(\alpha))\,(1-p_n(\alpha))}\right]=$$

$$= W_n\left(1-\gamma_n\left[|b|\,(1-\;p_n(\alpha))+(1-2c(\alpha))p_n(\alpha)\right]+O(\gamma_n^2)\right)\leq$$

$$\leq W_n\left(1-\gamma_n\min\left\{|b|\,;(1-2c(\alpha))\right\}+O(\gamma_n^2)\right) \tag{3.57}$$

Combining (3.47),(3.51) and (3.57) gives rise to the following inequality:

$$\mathbf{E}\left\{W_{n+1}/\mathcal{F}_{n-1}\bigwedge u_n = u(i)\right\} \overset{a.s.}{\leq} W_n\left(1-\gamma_n\min\left\{|b|\,;B\right\}+O(\gamma_n^2)\right) \tag{3.58}$$

where

$$B = 1-2c(\alpha)$$

From (3.58) and Robbins-Siegmund theorem [23] we obtain the convergence with probability one. Theorem is proved. ■

From this theorem it follows that a learning automaton using the Varshavskii-Vorontsova reinforcement scheme and operating in random stationary environments with nonbinary responses, belonging to the unit segment, selects asymptotically the optimal action $u_n = u(\alpha)$, i.e., its behaviour is asymptotically optimal.

Corollary 4. *Under the assumptions of this theorem the gain (correction factor) γ_n can be selected as a constant (satisfying the second condition of this theorem), i.e.,*

$$\gamma_n = \gamma$$

and then, the convergence rate is exponential and described by the formula

$$\mathbf{E}\{W_{n+1}\} \leq (N-1)(1-C)^n \tag{3.59}$$

where

$$C = \gamma \min\{|b|; (1-2c(\alpha))\}(1-\varepsilon) \in (0,1), \ 0 < \varepsilon \text{ small enough}$$

Proof.

follows directly from (3.58) after applying the operator of mathematical expectation. ■

The rate of the learning process is great if the parameter γ in the reinforcement scheme (3.46) is close to its possible upper value defined by the third condition of this theorem.

Corollary 5. *If under the assumptions of this theorem*

$$\gamma_n = \frac{\gamma}{t+a}, \ a < \gamma \in (0,1]$$

Then, the convergence rate is described by the formula

$$\mathbf{E}\{W_{n+1}\} \leq (N-1)\prod_{t=1}^{n}\left(1 - \frac{\gamma}{t+a}\min\{|b|; B\}(1+o(1))\right) < \\ < (N-1)\, n^{-C} \tag{3.60}$$

where

$$B = 1 - 2c(\alpha)$$

and

$$C = [\gamma - \varepsilon]\min\{|b|; (1-2c(\alpha))\}, \ 0 < \varepsilon \text{ small enough}$$

Proof.

This result follows directly from (3.58) after averaging and using the following inequalities

$$\prod_{t=1}^{n}\left(1-\frac{\gamma}{t+a}\min\{|b|;(1-2c(\alpha))\}(1+o(1))\right)\leq$$

$$\leq\prod_{t=1}^{n}\left(1-\frac{1}{t}[\gamma-\varepsilon]\min\{|b|;(1-2c(\alpha))\}\right)$$

and

$$\prod_{t=1}^{n}[1-t^{-1}C]\leq\exp\left\{-C\sum_{t=1}^{n}t^{-1}\right\}\leq\exp\{-C\ln n\}=n^{-C}$$

Corollary is proved. ■

In this case the convergence rate also increases if the parameter γ increases too.

A remark is in order here. We have seen that the Lyapunov function seems to be helpful in our analysis, and this will be reinforced as we proceed.

The next section deals with the normalization procedure and the optimization algorithm.

3.5 Normalization Procedure and Optimization

Let us now show how to apply the nonprojectional reinforcement schemes for solving the Stochastic Optimization Problem (3.5) on discrete sets using only the corresponding observations y_n (3.2) of the function $f(x)$ to be optimized. It is evident that we can not directly use these schemes for optimization purposes. In fact, **environment responses have to belong to the unit interval** $[0,1]$, **but the available observations do not obligatory satisfy this conditions.**

What to do? This is the central question of this book. The solution consists of introducing a mapping between the observations y_n and the automaton inputs ξ_n (environment responses). A procedure called *"normalization procedure"*, which establish the connection between the environment response ξ_n and the available observations y_n has been initially described in [14] in order to keep also the automaton input within the unit segment $[0,1]$.

Definition 1. *We will define the normalization procedure by the following relation:*

$$\xi_n = \widetilde{\xi}_n := \frac{\left[\widetilde{c}_n(i) - \min_j \widetilde{c}_{n-1}(j)\right]_+}{\max_k \left[\widetilde{c}_n(k) - \min_j \widetilde{c}_{n-1}(j)\right]_+ + 1}, \qquad u_n = u(i) \tag{3.61}$$

where

$$\widetilde{c}_n(i) := \frac{\sum_{t=1}^{n} y_t \chi(u_t = u(i))}{\sum_{t=1}^{n} \chi(u_t = u(i))}, \quad i = 1, ..., N \tag{3.62}$$

$$[x]_+ := \begin{cases} x & \text{if} \quad x \geq 0 \\ 0 & \text{if} \quad x < 0 \end{cases} \tag{3.63}$$

The following assumptions concerning the properties of the observation noise $\{w_n\}$ will be in force throughout of this book.

(H3) The conditional mathematical expectations of the observation noise w_n are equal to zero for any time $n = 1, 2, ...$, i.e.,

$$\mathbf{E}\left\{w_n / \mathcal{F}_{n-1} \bigwedge u_n = u(i)\right\} \overset{a.s.}{=} 0$$

(It means that $\{w_n\}$ is a sequence containing martingale-differences.)

(H4) The conditional variances of the observation noises exist and are uniformly bounded, i.e.,

$$\mathbf{E}\left\{w_n^2 / \mathcal{F}_{n-1} \bigwedge u_n = u(i)\right\} \overset{a.s.}{=} \sigma_n^2(i)$$

$$\max_i \sup_n \sigma_n^2(i) := \sigma^2 < \infty$$

The main properties of the normalization procedure (3.61) is presented in the following lemma.

Lemma 2. *Assume that assumptions (H3) and (H4) hold and suppose that any reinforcement scheme (nonprojectional or projectional) generates the sequence $\{p_n\}$ such that for any $i = 1, ..., N$*

$$\sum_{n=1}^{\infty} p_n(i) \overset{a.s.}{=} \infty \tag{3.64}$$

Then, the normalized environment response $\widetilde{\xi}_n$ (3.61) possesses the following properties:

- *The number of selections of each action is infinite, i.e.,*

$$\sum_{t=1}^{\infty} \chi(u_t = u(i)) \overset{a.s.}{=} \infty \quad \forall i = 1, ..., N \tag{3.65}$$

- *The random variable $\widetilde{c}_n(i)$ is asymptotically equal to the value of the function to be optimized for the corresponding point $x(i)$ belonging to the finite set $\mathcal{X}$, i.e.,*

$$\widetilde{c}_n(i) = f(x(i)) + o_\omega(1), \qquad \forall i = 1, ..., N \tag{3.66}$$

- *For the selected action $u_n = u(i)$ at time n, the normalized environment reaction $\widetilde{\xi}_n$ is asymptotically equal to $\Delta(i)$, i.e.,*

$$\widetilde{\xi}_n \overset{a.s.}{=} \Delta(i) + o_\omega(1) \in [0, 1), \quad u_n = u(i) \quad \forall i = 1, ..., N \tag{3.67}$$

  where

$$\Delta(i) = \frac{f(x(i)) - f(x(\alpha))}{\max_i [f(x(i)) - f(x(\alpha))] + 1} \in [0, 1) \tag{3.68}$$

- *For the optimal action $u_n = u(\alpha)$, the normalized environment reaction is asymptotically equal to 0, i.e.,*

$$\widetilde{\xi}_n \overset{a.s.}{=} o_\omega(1), \quad u_n = u(\alpha) \tag{3.69}$$

Proof.

1. (3.65) follows directly from assumption (3.64) and the Borel-Cantelli lemma [22].

2. Let us introduce the following sequence

$$\theta_n(i) := \widetilde{c}_n(i) - f(x(i)) = \frac{\sum_{t=1}^{n} [y_t - f(x(i)] \chi(u_t = u(i))}{\sum_{t=1}^{n} \chi(u_t = u(i))}$$

$$= \frac{\sum_{t=1}^{n} w_t \chi(u_t = u(i))}{\sum_{t=1}^{n} \chi(u_t = u(i))}, \quad i = 1, ..., N$$

   which leads to the following recurrent form for $\theta_n(i)$

$$\theta_n(i) = (1 - \lambda_n(i))\theta_{n-1}(i) + \lambda_n(i) w_n$$

where

$$\lambda_n(i) := \frac{\chi(u_n = u(i))}{\sum_{t=1}^{n} \chi(u_t = u(i))}$$

Taking into account assumptions **(H3)**,**(H4)** we derive

$$\mathbf{E}\left\{ (\theta_n(i))^2 / \mathcal{F}_{n-1} \bigwedge u_n = u(i) \right\} \stackrel{a.s.}{=}$$

$$\stackrel{a.s.}{=} (1 - \lambda_n(i))^2 (\theta_{n-1}(i))^2 + (\lambda_n(i))^2 \sigma_n^2(i) \leq$$
$$\leq (1 - \lambda_n(i))^2 (\theta_{n-1}(i))^2 + (\lambda_n(i))^2 \sigma^2 \qquad (3.70)$$

It is easy to see that

$$\sum_{t=1}^{n} \lambda_n(i) \stackrel{a.s.}{=} \infty, \ \sum_{t=1}^{n} \lambda_n^2(i) \stackrel{a.s.}{<} \infty$$

In view of this facts and Robbins-Siegmund theorem [23], it follows $\theta_n(i) \underset{n\to\infty}{\stackrel{a.s.}{\to}} 0$ and hence

$$\lim_{n\to\infty} \widetilde{c}_n(i) = f(x(i)) \ \forall i = 1, ..., N$$

So, (3.66) is proved.

3. (3.67) follows directly from (3.66) and assumption **(H3)**.
4. (3.69) is the consequence of (3.67) and (3.68). ■

Corollary 6. *If for some reinforcement scheme the following inequality holds*

$$\frac{1}{n}\sum_{t=1}^{n} p_t(i) \geq O\left(\frac{1}{n^\tau}\right), \ \tau \in \left(0, \frac{1}{2}\right) \qquad (3.71)$$

Then, for any small positive ε *we have*

$$\widetilde{c}_n(i) - f(x(i)) \stackrel{a.s.}{=} o_\omega\left(\frac{1}{n^{1/2-\tau-\varepsilon}}\right) \qquad (3.72)$$

and, as a result, for large enough $n \geq n_0(\omega)$, *we obtain*

$$\widetilde{\xi}_n \stackrel{a.s.}{=} \Delta(i) + o_\omega\left(\frac{1}{n^{1/2-\tau-\varepsilon}}\right) \in [0,1), \quad u_n = u(i) \quad \forall i = 1, ..., N \qquad (3.73)$$

$$\mathbf{E}\left\{ \left(\widetilde{\xi}_n - \Delta(i)\right)^2 \mid u_n = u(i) \right\} = o\left(\frac{1}{n^{1/2-\tau}}\right) \qquad (3.74)$$

and

$$\mathbf{E}\left\{ \widetilde{\xi}_n \mid u_n = u(i) \right\} = \Delta(i) + o\left(\frac{1}{n^{1-2\tau}}\right) \qquad (3.75)$$

Proof.

Let us notice that according to the strong law of large numbers [21]-[22] and lemma A.13a in [14], it follows

$$\frac{1}{n}\sum_{t=1}^{n}\left[\chi(u_t=u(i))-p_t(i)\right]=o_\omega(n^{-(1/2-\varepsilon)})$$

and for

$$\lambda_n(i)=\frac{\chi(u_n=u(i))}{\sum\limits_{t=1}^{n}\chi(u_t=u(i))}=\frac{\chi(u_n=u(i))}{n\left[\frac{1}{n}\sum\limits_{t=1}^{n}p_t(i)+o_\omega(n^{-(1/2-\varepsilon)})\right]}$$

we obtain two estimations for large enough $n\geq n_0(\omega)$:

$$\lambda_n(i)\geq\frac{\chi(u_n=u(i))}{n\left[1+o_\omega(n^{-(1/2-\varepsilon)})\right]}\geq\frac{\chi(u_n=u(i))\left[1-\varepsilon\right]}{n}$$

and

$$\lambda_n(i)\leq\frac{\chi(u_n=u(i))}{n\left[O\left(\frac{1}{n^\tau}\right)+o_\omega(n^{-(1/2-\varepsilon)})\right]}\leq$$
$$\leq\chi(u_n=u(i))\left[1+o_\omega(1)\right]O\left(n^{-1+\tau}\right)\leq$$
$$\leq\chi(u_n=u(i))\left[1+\varepsilon\right]O\left(n^{-1+\tau}\right)$$

From the last inequality it follows also that

$$\lambda_n^2(i)\leq\chi(u_n=u(i))\left[1+\varepsilon\right]O\left(n^{-2[1-\tau]}\right)$$

1. In view of lemma A.14 given in [14], from (3.70) we derive

$$\widetilde{c}_n(i)-f(x(i))\overset{a.s.}{=}o_\omega\left(\frac{1}{n^{1/2-\tau-\varepsilon}}\right)$$

from which (3.72) and (3.73) follow. Averaging (3.70) and using lemma A.1 in [14] we obtain (3.74), from which according to Jensen's inequality and the following relations

$$\left|\mathbf{E}\left\{\widetilde{\xi}_n-\Delta(i)\right|u_n=u(i)\right\}\right|\leq\mathbf{E}\left\{\left|\widetilde{\xi}_n-\Delta(i)\right|\mid u_n=u(i)\right\}\leq$$

$$\leq\sqrt{\mathbf{E}\left\{\left(\widetilde{\xi}_n-\Delta(i)\right)^2\mid u_n=u(i)\right\}}=o\left(\frac{1}{n^{1/2-\tau}}\right)$$

we derive (3.75). Corollary is proved. ■

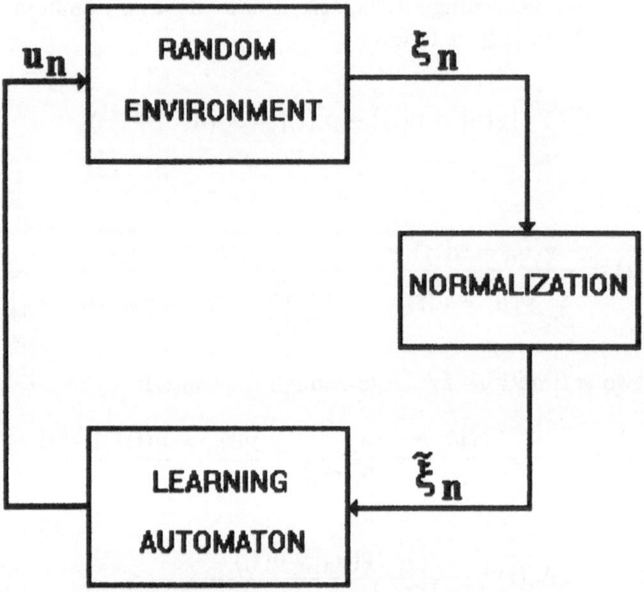

FIGURE 3.1. Feedback connection between automaton and environment with normalized response.

In the following we shall be concerned with the behaviour of learning automata with normalized environment response and their use for optimization purposes. The probability distribution p_n will be adjusted using the previous reinforcement schemes, where ξ_n is constructed according to the procedure (3.61) (see Figure 3.1).

This Figure represents the feedback connection between the automaton and the environment for which the response is normalized.

Taking into account the optimization problem to be solved, the environment response is constructed as it shown by Figure 3.2. The admissible compact region X is quantified into a set of subsets, and the loss function is calculated using the noisy observations (realizations) of the function to be optimized.

The convergence analysis of the behaviour of learning automata using the reinforcement schemes described before and the normalization procedure is given in the next section.

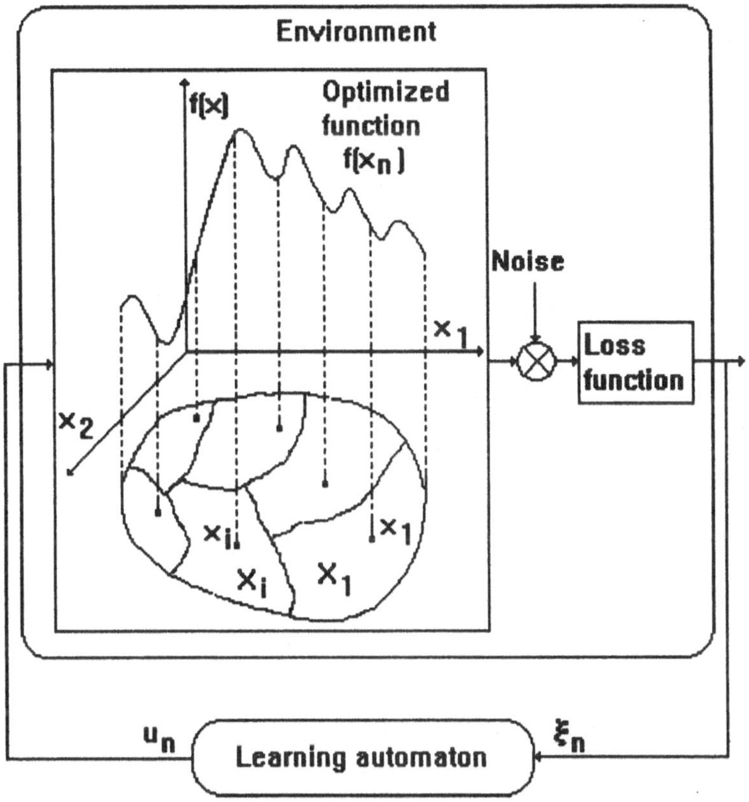

FIGURE 3.2. Multimodal searching technique based on learning automaton.

3.6 Reinforcement Schemes with Normalized Environment response

The aim of the following subsections is to examine and analyze the properties of S-model learning automata using respectively Bush-Mosteller [18], Shapiro-Narendra [19] and Varshavskii-Vorontsova [20] reinforcement schemes will be given in the following subsections. The environment response is normalized according to the procedure (3.61)-(3.63).

3.6.1 Bush-Mosteller reinforcement scheme with normalization procedure

The correction factor was considered constant ($\gamma_n = \gamma = const.$) in the original version of these reinforcement schemes. In this study, we assume that the correction factor γ_n is time-varying and is selected according to the following formula:

$$\gamma_n = \frac{\gamma}{n+a}, \ \gamma \in (0,1), a > \gamma \tag{3.76}$$

The initial probabilities are assumed to be strictly positive ($p_1(i) > 0 \quad \forall i = 1, ..., N$).

The Bush-Mosteller scheme [18]-[14] is described by:

$$p_{n+1} = p_n + \gamma_n \left[e(u_n) - p_n + \tilde{\xi}_n (e^N - N e(u_n))/(N-1) \right] \tag{3.77}$$

with

$$\begin{aligned} \gamma_n &= \frac{\gamma}{n+a}, \ \gamma \in (0,1), a > \gamma, \quad \tilde{\xi}_n \in [0,1) \\ e(u_n) &= (\underbrace{0, ..., 0, 1}_{i}, 0, ..., 0)^T, \ u_n = u(i) \\ e^N &= (1, ..., 1)^T \in R^N \end{aligned}$$

Theorem 6. *For the Bush-Mosteller scheme (3.77), condition (3.64) is satisfied, and if assumptions (H3) and (H4) hold, and the optimal action is single, i.e.,*

$$\min_{i \neq \alpha} \Delta(i) := \Delta^* > 0 \tag{3.78}$$

Then, the automaton selects asymptotically the global optimal point $x(\alpha)$ and the loss function Φ_n tends to its minimal possible value $f(x(\alpha))$, with probability one.

Proof.

Let us consider the following estimation:

$$\begin{aligned} p_{n+1}(i) = p_n(i) + \gamma_n [\chi(u_n = u(i)) - p_n(i) + \\ + \tilde{\xi}_n [1 - N\chi(u_n = u(i))]/(N-1)] = p_n(i)\ (1-\gamma_n) + \\ + \frac{\gamma_n}{N-1} \left\{ \tilde{\xi}_n + \chi(u_n = u(i)) \left[N(1-\tilde{\xi}_n) - 1 \right] \right\} \geq \\ \geq p_n(i)\ (1-\gamma_n) \geq \cdots \geq p_1(i) \prod_{t=1}^{n} (1-\gamma_t) \end{aligned} \tag{3.79}$$

From lemma A.4 [14], it follows that

$$\prod_{t=1}^{n}(1-\gamma_t) \geq \begin{cases} \left(\frac{a-\gamma}{n+a}\right)^{\gamma}, & a > \gamma \in (0,1) \\ \frac{a}{n+a}, & \gamma = 1, \; a > 0 \end{cases} \tag{3.80}$$

Substituting (3.80) into (3.79) leads to the desired result (*3.64*). Notice also that from (3.80) and (3.79) it follows that in (3.72)

$$\tau = \gamma \in (0, 1/2)$$

and hence in (3.73) we have

$$\widetilde{c}_n(i) - f(x(i)) \overset{a.s.}{=} o_{\omega}\left(\frac{1}{n^{1/2-\gamma-\varepsilon}}\right) \tag{3.81}$$

Performing similar calculation as in (3.81) and (3.25), and taking into account the reinforcement scheme (3.77), we can obtain the following inequality for s_n which is similar to (3.29):

$$s_n \overset{a.s.}{\leq} n^{-2(1-\gamma)} C_{n1} \left(1 + o(1)\right) W_n, \; C_{n1} := \frac{\gamma^2 C_n}{(a-\gamma)^{2\gamma} (N-1)^2} \tag{3.82}$$

with

$$C_n = \sum_{i \neq \alpha}^{N} \mathbf{E}\left\{\left(\widetilde{\xi}_n - \Delta(i)\right)^2 | u_n = u(i)\right\} = o\left(\frac{1}{n^{1-2\gamma}}\right) \tag{3.83}$$

Substituting (3.83) into (3.82) we get:

$$s_n \overset{a.s.}{\leq} o(n^{-3+4\gamma}) W_n$$

Calculating the first term in (3.24) we derive:

$$\sum_{i=1}^{N}\left(\frac{1}{\mathbf{E}\left\{p_{n+1}(\alpha)/\mathcal{F}_{n-1} \bigwedge u_n = u(i)\right\}} - 1\right) p_n(i) =$$

$$= \sum_{i \neq \alpha}^{N}\left(\frac{1}{\mathbf{E}\left\{p_{n+1}(\alpha)/\mathcal{F}_{n-1} \bigwedge u_n = u(i)\right\}} - 1\right) p_n(i) +$$

$$+ \left(\frac{1}{\mathbf{E}\left\{p_{n+1}(\alpha)/\mathcal{F}_{n-1} \bigwedge u_n = u(\alpha)\right\}} - 1\right) p_n(\alpha) =$$

$$= \sum_{i \neq \alpha}^{N}\left(\frac{1}{\left[p_n(\alpha)(1-\gamma_n) + \frac{\gamma_n}{N-1}\left(\Delta(i) + o(n^{-1/2+\gamma})\right)\right]} - 1\right) p_n(i) +$$

$$+\left(\frac{1}{p_n(\alpha)(1-\gamma_n)+\gamma_n\left[1+o(n^{-1/2+\gamma})\right]}-1\right)p_n(\alpha)\leq$$

$$\leq\left(\frac{1}{p_n(\alpha)(1-\gamma_n)+\frac{\gamma_n}{N-1}\left(\Delta^*+o(n^{-1/2+\gamma})\right)}-1\right)(1-p_n(\alpha))+$$

$$+\frac{(1-p_n(\alpha))(1-\gamma_n)+\gamma_n o(n^{-1/2+\gamma})}{p_n(\alpha)+\gamma_n\left(1-p_n(\alpha)+o(n^{-1/2+\gamma})\right)}p_n(\alpha)=$$

Notice that

$$\frac{1}{N-1}\Delta^*<1$$

we derive

$$\sum_{i=1}^{N}\left(\frac{1}{\mathbf{E}\left\{p_{n+1}(\alpha)/\mathcal{F}_{n-1}\bigwedge u_n=u(i)\right\}}-1\right)p_n(i)\overset{a.s.}{\leq}$$

$$\leq\frac{(1-p_n(\alpha))\left(1-p_n(\alpha)+\gamma_n\left[p_n(\alpha)-\frac{1}{N-1}\left(\Delta^*+o(n^{-1/2+\gamma})\right)\right]\right)}{p_n(\alpha)+\gamma_n\left(\frac{1}{N-1}\Delta^*-p_n(\alpha)+o(n^{-1/2+\gamma})\right)}+$$

$$+\frac{(1-p_n(\alpha))(1-\gamma_n)+\gamma_n o(n^{-1/2+\gamma})}{p_n(\alpha)+\gamma_n\left(1-p_n(\alpha)+o(n^{-1/2+\gamma})\right)}p_n(\alpha)\leq$$

$$\leq\frac{(1-p_n(\alpha))\left(1-p_n(\alpha)+\gamma_n\left[p_n(\alpha)-\frac{1}{N-1}\left(\Delta^*+o(n^{-1/2+\gamma})\right)\right]\right)}{p_n(\alpha)+\gamma_n\left(\frac{1}{N-1}\Delta^*-p_n(\alpha)+o(n^{-1/2+\gamma})\right)}+$$

$$+\frac{(1-p_n(\alpha))(1-\gamma_n)+\gamma_n o(n^{-1/2+\gamma})}{p_n(\alpha)+\gamma_n\left(\frac{1}{N-1}\Delta^*-p_n(\alpha)+o(n^{-1/2+\gamma})\right)}p_n(\alpha)=$$

$$=\frac{(1-p_n(\alpha))\left(1-\gamma_n\frac{1}{N-1}\Delta^*\right)+\gamma_n o(n^{-1/2+\gamma})}{p_n(\alpha)+\gamma_n\left(\frac{1}{N-1}\Delta^*-p_n(\alpha)+o(n^{-1/2+\gamma})\right)}=$$

$$=W_n\left(1-\gamma_n\frac{1}{N-1}\Delta^*\right)\frac{1}{1-\gamma_n+\frac{\gamma_n}{p_n(\alpha)}\left(\frac{1}{N-1}\Delta^*+o(n^{-1/2+\gamma})\right)}+$$

$$+\frac{\gamma_n}{p_n(\alpha)}o(n^{-1/2+\gamma})\frac{1}{1-\gamma_n+\frac{\gamma_n}{p_n(\alpha)}\left(\frac{1}{N-1}\Delta^*+o(n^{-1/2+\gamma})\right)}=$$

Taking into account (3.22), (3.23), we conclude that

$$\gamma_n\leq\frac{\gamma_n}{p_n(\alpha)}\leq\frac{C_2(1+o(1))}{n^{1-\gamma}},\ C_2:=\frac{\gamma\Delta^*}{(N-1)p_1(\alpha)(a-\gamma)^{\gamma}}$$

and hence,

$$\sum_{i=1}^{N}\left(\frac{1}{\mathbf{E}\{p_{n+1}(\alpha)/\mathcal{F}_{n-1}\bigwedge u_n=u(i)\}}-1\right)p_n(i)\overset{a.s.}{\leq}$$

$$\leq W_n\left(1-\gamma_n\frac{1}{N-1}\Delta^*\right)\frac{1}{1+\frac{C_2[1+o(1)]}{n^{1-\gamma}}}+o(n^{-3/2+2\gamma})=$$

$$=W_n\left(1-\gamma_n\frac{1}{N-1}\Delta^*\right)\left(1-\frac{C_2\left[1+o(1)\right]}{n^{1-\gamma}}\right)+o(n^{-3/2+2\gamma})=$$

$$=W_n\left(1-\frac{C_2\left[1+o(1)\right]}{n^{1-\gamma}}\right)+o(n^{-3/2+2\gamma}) \tag{3.84}$$

for $n\geq n_0(\omega)$, $n_0(\omega)\overset{a.s.}{<}\infty$.
Combining (3.81), (3.82) and (3.84) we obtain

$$\mathbf{E}\{W_{n+1}/\mathcal{F}_{n-1}\}\overset{a.s.}{\leq}W_n\left(1-\frac{C_2\left[1+o(1)\right]}{n^{1-\gamma}}+o(n^{-3+4\gamma})\right)+o(n^{-3/2+2\gamma})=$$

$$=W_n\left(1-\frac{C_2\left[1+o(1)\right]}{n^{1-\gamma}}\right)+o(n^{-3/2+2\gamma}) \tag{3.85}$$

In view of Robbins-Siegmund theorem [23], it follows

$$W_n\underset{n\to\infty}{\overset{a.s.}{\to}}0,\ p_n(\alpha)\underset{n\to\infty}{\overset{a.s.}{\to}}1$$

Notice now that the hypotheses (**H1**) and (**H2**) follow from (**H3**) and (**H4**), and hence, according to the third theorem of this chapter, we obtain

$$\Phi_n\underset{n\to\infty}{\overset{a.s.}{\to}}f(x(\alpha))$$

Theorem is proved. ■

This theorem shows that, a learning automaton using the Bush-Mosteller reinforcement scheme with the normalization procedure described above, selects asymptotically the optimal action, i.e., the optimal point $x(\alpha)$ is asymptotically selected.

We still have to show how to choose the design parameters of the optimization algorithm such that to increase the speed of convergence.

The next corollary gives the estimation of the rate of optimization.

Corollary 7. *Under the assumptions of this theorem it follows (convergence rate):*

$$W_n\overset{a.s.}{=}o\left(\frac{1}{n^{\nu}}\right)$$

where

$$0<\nu<\frac{1}{2}-2\gamma$$

Proof.

In view of lemma A.14 in [14] for

$$\nu_n := n^{\nu},\ u_n := W_n,\ \alpha_n := \frac{C_2\,[1+o(1)]}{n^{1-\gamma}},\ \beta_n := o(n^{-3/2+2\gamma})$$

From (3.85), we obtain

$$\frac{3}{2} - 2\gamma - \nu > 1$$

Corollary is proved. ■

The next subsection deals with the analysis of Shapiro-Narendra reinforcement scheme using the normalization procedure.

3.6.2 Shapiro-Narendra reinforcement scheme with normalization procedure

The Shapiro-Narendra scheme [19]-[14] is described by:

$$p_{n+1} = p_n + \gamma_n(1-\widetilde{\xi}_n)\,[e(u_n) - p_n] \tag{3.86}$$

$$\gamma_n, \widetilde{\xi}_n \in [0,1]$$

We shall analyse its behaviour when the environment responses $\widetilde{\xi}_n$ are constructed on the basis of the realizations of the function to be optimized (Figure 3.2) and the normalization procedure (Figure 3.1).

Theorem 7. *For the Shapiro-Narendra scheme [19] (3.86), assume that the optimal point $x(\alpha)$ is single and suppose that assumptions (H3), (H4) hold. In addition, suppose that the correction factor satisfies the following condition:*

1.

$$\sum_{n=1}^{\infty} \gamma_n = \infty,\ \sum_{n=1}^{\infty}\prod_{t=1}^{n}(1-\gamma_t) = \infty, \tag{3.87}$$

$$\sum_{n=1}^{\infty}\left[\frac{\gamma_n}{n^{1-2\tau}}\prod_{t=1}^{n}(1-\gamma_t)^{-1} + \gamma_n^2\prod_{t=1}^{n}(1-\gamma_t)^{-2}\right] < \infty$$

2.

$$\lim_{n\to\infty}\ \gamma_n\prod_{t=1}^{n}(1-\gamma_t)^{-1} := c < p_1(\alpha)\Delta^* \tag{3.88}$$

3.
$$\frac{1}{n}\sum_{t=1}^{n}\prod_{s=1}^{t}(1-\gamma_s) \geq O(\frac{1}{n^\tau}),\ \tau \in \left(0, \frac{1}{2}\right) \tag{3.89}$$

Then, this reinforcement scheme select asymptotically the optimal point $x(\alpha)$, i.e.,

$$p_n(\alpha) \underset{n\to\infty}{\overset{a.s.}{\to}} 1$$

and the loss function Φ_n tends to its minimal possible value $f(x(\alpha))$, with probability one.

Proof.

The proof is similar in structure to the proof of the previous theorem. Let us estimate the lower bounds of the probabilities $p_{n+1}(i)$:

$$p_{n+1}(i) = p_n(i) + \gamma_n(1-\widetilde{\xi}_n)\left[\chi(u_n = u(i)) - p_n(i)\right] = \tag{3.90}$$

$$= p_n(i)\left[1-\gamma_n(1-\widetilde{\xi}_n)\right] + \gamma_n(1-\widetilde{\xi}_n)\chi(u_n = u(i)) \geq$$

$$\geq p_n(i)\ (1-\gamma_n) \geq \cdots \geq p_1(i)\prod_{t=1}^{n}(1-\gamma_t)$$

From (3.87) and (3.90) it follows that

$$\sum_{n=1}^{\infty} p_n(i) = \infty,\ \forall i = 1, ..., N$$

As a result, according to the Borel-Cantelli lemma [22], we obtain

$$\sum_{n=1}^{\infty} \chi\left(u_n = u(i)\right) = \infty,\ \forall i = 1, ..., N \tag{3.91}$$

Let us consider again the following Lyapunov function

$$W_n := \frac{1-p_n(\alpha)}{p_n(\alpha)}$$

Notice that in view of assumption (3.89), the relations (3.72) and (3.75) are fulfilled, i.e.,

$$\widetilde{c}_n(i) - f(x(i)) \overset{a.s.}{=} o_\omega(\frac{1}{n^{1/2-\tau-\varepsilon}}) \tag{3.92}$$

$$\mathbf{E}\left\{\widetilde{\xi}_n \middle|\ u_n = u(i)\right\} = \Delta(i) + o(\frac{1}{n^{1-2\tau}})$$

Taking into account the Shapiro Narendra scheme (3.86) and (3.92), it follows:

$$\mathbf{E}\left\{W_{n+1}/\mathcal{F}_n\right\} \overset{a.s.}{=} \sum_{i=1}^{N} \mathbf{E}\left\{W_{n+1}/\mathcal{F}_{n-1} \bigwedge u_n = u(i)\right\} p_n(i) =$$

$$= \sum_{i\neq\alpha}^{N} \mathbf{E}\left\{\left(\frac{1}{p_n(\alpha) - \gamma_n(1-\widetilde{\xi}_n)\, p_n(\alpha)} - 1\right)/\mathcal{F}_{n-1} \bigwedge u_n = u(i)\right\} p_n(i) +$$

$$+\mathbf{E}\left\{\left(\frac{1}{p_n(\alpha) + \gamma_n(1-\widetilde{\xi}_n)\,(1 - p_n(\alpha))} - 1\right)/\mathcal{F}_{n-1} \bigwedge u_n = u(i)\right\} \times$$

$$\times p_n(\alpha) =$$

$$= \sum_{i\neq\alpha}^{N} \mathbf{E}\left\{\left(\frac{1}{p_n(\alpha)\,(1-\gamma_n) + \gamma_n\widetilde{\xi}_n\, p_n(\alpha)} - 1\right)/\mathcal{F}_{n-1} \bigwedge u_n = u(i)\right\} p_n(i) +$$

$$+\mathbf{E}\left\{\left(\frac{1}{p_n(\alpha)\,(1-\gamma_n) + \gamma_n - \gamma_n\widetilde{\xi}_n\,(1 - p_n(\alpha))} - 1\right)/\right.$$

$$\left./\ \mathcal{F}_{n-1} \bigwedge u_n = u(i)\right\} \times p_n(\alpha) =$$

$$= \sum_{i\neq\alpha}^{N} \mathbf{E}\left\{\left(\frac{1 - \frac{\gamma_n\widetilde{\xi}_n}{(1-\gamma_n)}}{p_n(\alpha)\,(1-\gamma_n)} - 1\right)/\mathcal{F}_{n-1} \bigwedge u_n = u(i)\right\} p_n(i) +$$

$$+\mathbf{E}\left\{\left(\frac{1 + \frac{\gamma_n\widetilde{\xi}_n(1-p_n(\alpha))}{p_n(\alpha)(1-\gamma_n)+\gamma_n}}{p_n(\alpha)\,(1-\gamma_n) + \gamma_n} - 1\right)/\mathcal{F}_{n-1} \bigwedge u_n = u(i)\right\} p_n(\alpha) +$$

$$+O\left(\frac{\gamma_n^2}{p_n^2(\alpha)}\right) =$$

$$= \sum_{i\neq\alpha}^{N} \left(\frac{1 - \frac{\gamma_n}{(1-\gamma_n)}\left[\Delta(i) + o(\frac{1}{n^{1-2\tau}})\right]}{p_n(\alpha)\,(1-\gamma_n)} - 1\right) p_n(i) +$$

$$+\left(\frac{1 + \frac{\gamma_n(1-p_n(\alpha))}{p_n(\alpha)(1-\gamma_n)+\gamma_n} o(\frac{1}{n^{1-2\tau}})}{p_n(\alpha)\,(1-\gamma_n) + \gamma_n} - 1\right) p_n(\alpha) + O\left(\frac{\gamma_n^2}{p_n^2(\alpha)}\right) \leq$$

$$\leq \left(\frac{1 - \frac{\gamma_n}{(1-\gamma_n)}\left[\Delta^* + o(\frac{1}{n^{1-2\tau}})\right]}{p_n(\alpha)\,(1-\gamma_n)} - 1\right)(1 - p_n(\alpha)) +$$

$$+\left(\frac{1}{p_n(\alpha)\,(1-\gamma_n) + \gamma_n} - 1\right) p_n(\alpha) + \frac{\gamma_n}{p_n(\alpha)} o(\frac{1}{n^{1-2\tau}}) + O\left(\frac{\gamma_n^2}{p_n^2(\alpha)}\right) =$$

$$= \left(\frac{1 - \frac{\gamma_n}{(1-\gamma_n)} \Delta^*}{p_n(\alpha)(1-\gamma_n)} - 1 \right) (1 - p_n(\alpha)) + \left(\frac{1}{p_n(\alpha)(1-\gamma_n) + \gamma_n} - 1 \right) p_n(\alpha) +$$

$$+ \frac{\gamma_n}{p_n(\alpha)} o\left(\frac{1}{n^{1-2\tau}}\right) + O\left(\frac{\gamma_n^2}{p_n^2(\alpha)}\right)$$

Inequality (3.90) gives

$$\gamma_n p_n^{-1}(\alpha) \le p_1^{-1}(\alpha) \gamma_n \prod_{t=1}^{n} (1 - \gamma_t)^{-1} \tag{3.93}$$

and hence,

$$\mathbf{E}\{W_{n+1}/\mathcal{F}_n\} \overset{a.s.}{\le}$$

$$\overset{a.s.}{\le} \left(1 - \frac{\gamma_n}{(1-\gamma_n)} \Delta^* - p_n(\alpha)(1-\gamma_n)\right) \frac{1 - p_n(\alpha)}{p_n(\alpha)(1-\gamma_n)} +$$

$$+ \frac{(1 - p_n(\alpha))(1-\gamma_n)}{p_n(\alpha)(1-\gamma_n) + \gamma_n} p_n(\alpha) +$$

$$+ o\left(\frac{1}{n^{1-2\tau}}\right) \gamma_n \prod_{t=1}^{n} (1-\gamma_t)^{-1} + O\left(\gamma_n^2 \prod_{t=1}^{n} (1-\gamma_t)^{-2}\right) =$$

$$= W_n \left[(1 - \gamma_n \Delta^* - p_n(\alpha)(1-\gamma_n))(1+\gamma_n) + \right.$$

$$\left. + \frac{(1-\gamma_n) p_n^2(\alpha)}{p_n(\alpha)(1-\gamma_n) + \gamma_n} + O(\gamma_n^2) \right] +$$

$$+ o\left(\frac{1}{n^{1-2\tau}}\right) \gamma_n \prod_{t=1}^{n} (1-\gamma_t)^{-1} + O\left(\gamma_n^2 \prod_{t=1}^{n} (1-\gamma_t)^{-2}\right) =$$

$$= W_n \left[(1 - \gamma_n \Delta^*)(1+\gamma_n) - \frac{\gamma_n p_n(\alpha)}{p_n(\alpha)(1-\gamma_n) + \gamma_n} + O(\gamma_n^2) \right] +$$

$$+ o\left(\frac{1}{n^{1-2\tau}}\right) \gamma_n \prod_{t=1}^{n} (1-\gamma_t)^{-1} + O\left(\gamma_n^2 \prod_{t=1}^{n} (1-\gamma_t)^{-2}\right) =$$

$$= W_n \left[1 + \gamma_n (1 - \Delta^*) - \gamma_n \frac{1}{1 + \gamma_n \left(p_n^{-1}(\alpha) - 1\right)} + O(\gamma_n^2) \right] +$$

$$+ o\left(\frac{1}{n^{1-2\tau}}\right) \gamma_n \prod_{t=1}^{n} (1-\gamma_t)^{-1} + O\left(\gamma_n^2 \prod_{t=1}^{n} (1-\gamma_t)^{-2}\right) =$$

$$= W_n \left[1 - \gamma_n \left(\Delta^* - \gamma_n p_n^{-1}(\alpha)\right) + O(\gamma_n^2 p_n^{-1}(\alpha)) \right] +$$

$$+ o\left(\frac{1}{n^{1-2\tau}}\right) \gamma_n \prod_{t=1}^{n} (1-\gamma_t)^{-1} + O\left(\gamma_n^2 \prod_{t=1}^{n} (1-\gamma_t)^{-2}\right)$$

Using again inequality (3.90) and the assumptions of this theorem we derive:

$$\mathbf{E}\{W_{n+1}/\mathcal{F}_n\} \overset{a.s.}{\leq} W_n \left[1-\gamma_n\left(\Delta^* - p_1^{-1}(\alpha)\gamma_n \prod_{t=1}^{n}(1-\gamma_t)^{-1}\right) + \right.$$

$$\left. +O\left(\gamma_n^2 \prod_{t=1}^{n}(1-\gamma_t)^{-2}\right)\right] +$$

$$+o(\frac{1}{n^{1-2\tau}})\gamma_n \prod_{t=1}^{n}(1-\gamma_t)^{-1} + O\left(\gamma_n^2 \prod_{t=1}^{n}(1-\gamma_t)^{-2}\right) =$$

$$= W_n\left[1-\gamma_n\left(\Delta^* - p_1^{-1}(\alpha)c + o(1)\right) + O\left(\gamma_n^2 \prod_{t=1}^{n}(1-\gamma_t)^{-2}\right)\right] +$$

$$+o(\frac{1}{n^{1-2\tau}})\gamma_n \prod_{t=1}^{n}(1-\gamma_t)^{-1} + O\left(\gamma_n^2 \prod_{t=1}^{n}(1-\gamma_t)^{-2}\right) \tag{3.94}$$

Taking into account assumption (3.88) of this theorem $(\Delta^* - p_1^{-1}(\alpha)c > 0)$, assumption (3.87) and Robbins-Siegmund theorem [23], we obtain

$$W_n \underset{n\to\infty}{\overset{a.s.}{\to}} 0, \quad p_n(\alpha) \underset{n\to\infty}{\overset{a.s.}{\to}} 1$$

Theorem is proved. ■

The following corollary deals with the constraints associated with the correction factor γ_n.

Corollary 8. *For the correction factor (3.76)*

$$\gamma_n = \frac{\gamma}{n+a}, \ \gamma \in (0,1), a > \gamma$$

assumption (3.87), (3.88) and (3.89) will be satisfied if

$$\lim_{n\to\infty} \gamma_n \prod_{t=1}^{n}(1-\gamma_t)^{-1} = c = 0, \ \tau = \gamma$$

and:

1. the rate of optimization will be given by

$$n^{\nu} W_n \underset{n\to\infty}{\overset{a.s.}{\to}} 0 \tag{3.95}$$

where

$$\nu < \min\{1-3\gamma; \gamma\Delta^*\} := \nu(\gamma) \leq \nu^*$$

2. the maximum optimization rate ν^* is equal to

$$\max_{\gamma} \nu(\gamma) = \nu^* = \frac{\Delta^*}{\Delta^* + 3} \tag{3.96}$$

and is reached for

$$\gamma^* = \frac{1}{\Delta^* + 3} \tag{3.97}$$

Proof.

follows directly from inequality (3.94), which can be rewritten as follows:

$$\mathbf{E}\{W_{n+1}/\mathcal{F}_n\} \overset{a.s.}{\leq} W_n \left[1 - \frac{\gamma}{n}(\Delta^* + o(1))\right] + o(\frac{1}{n^{2-3\gamma}}) \tag{3.98}$$

Applying then lemma A.14 in [14] for

$$u_n := W_n,\ \alpha_n := \frac{\gamma}{n}(\Delta^* + o(1)),\ \beta_n := o(\frac{1}{n^{2-3\gamma}}),\ \nu_n := n^{\nu}$$

we obtain the desired result.

Corollary is proved. ■

These statements show that, a learning automaton using the Shapiro-Narendra reinforcement scheme with the normalization procedure described above, has an asymptotically optimal behaviour and the global optimization objective is achieved with an ε-accuracy.

The Varshavskii-Vorontsova reinforcement scheme will be considered in the following.

3.6.3 Varshavskii-Vorontsova Reinforcement Scheme with Normalization Procedure

The Varshavskii-Vorontsova scheme [20]-[14] is described by:

$$p_{n+1} = p_n + \gamma_n p_n^T\, e(u_n)(1 - 2\tilde{\xi}_n)\, [e(u_n) - p_n] \tag{3.99}$$

This scheme belongs to the class of nonlinear reinforcement schemes. We shall analyse its possibilities to solve the stochastic optimization problems on discrete sets. The following theorem cover all the specific properties of the Varshavskii-Vorontsova scheme.

Theorem 8. *For the Varshavskii-Vorontsova scheme (3.99), assume that the optimal point $x(\alpha)$ is single and suppose that assumptions (H3), (H4) hold. In addition, suppose that the correction factor satisfies the following condition:*

1.

$$\sum_{n=1}^{\infty} \gamma_n = \infty, \sum_{n=1}^{\infty} \prod_{t=1}^{n} (1-\gamma_t) = \infty, \tag{3.100}$$

$$\sum_{n=1}^{\infty} \left[\left(n^{-1+2\tau} + \gamma_n \right) \gamma_n \prod_{t=1}^{n} (1-\gamma_t)^{-1} \right] < \infty$$

2.

$$b := \left(\sum_{i \neq \alpha}^{N} [1 - 2\Delta(i)]^{-1} \right)^{-1} < 0 \tag{3.101}$$

3.

$$\frac{1}{n} \sum_{t=1}^{n} \prod_{s=1}^{t} (1-\gamma_s) \geq O(\frac{1}{n^\tau}), \; \tau \in \left(0, \frac{1}{2}\right) \tag{3.102}$$

Then, the optimal point $x(\alpha)$ is asymptotically selected, i.e.,

$$p_n(\alpha) \overset{a.s.}{\underset{n \to \infty}{\to}} 1$$

and the loss function Φ_n tends to its minimal possible value $f(x(\alpha))$, with probability one.

Proof.
Let us estimate the lower bounds of the probabilities $p_{n+1}(i)$:

$$p_{n+1}(i) = p_n(i) + \gamma_n p_n^T \; e(u_n)(1 - 2\tilde{\xi}_n)[\chi(u_n = u(i)) - p_n(i)] =$$

$$= p_n(i) \left[1 - \gamma_n p_n^T \; e(u_n)(1 - 2\tilde{\xi}_n) \right] + \gamma_n p_n^T \; e(u_n) \chi(u_n = u(i)) \geq$$

$$\geq p_n(i) \left[1 - \gamma_n p_n^T \; e(u_n)(1 - 2\tilde{\xi}_n) \right] \geq$$

$$\geq p_n(i) \left[1 - \gamma_n p_n^T \; e(u_n) \right] \geq$$

$$\geq p_n(i) \; (1 - \gamma_n) \geq \cdots \geq p_1(i) \prod_{t=1}^{n} (1 - \gamma_t) \tag{3.103}$$

From (3.100) and (3.103) it follows that

$$\sum_{n=1}^{\infty} p_n(i) = \infty, \; \forall i = 1, ..., N$$

As a result, and according to the Borel-Cantelli lemma [22], we obtain

$$\sum_{n=1}^{\infty} \chi\left(u_n = u(i)\right) = \infty, \ \forall i = 1, ..., N \tag{3.104}$$

Let us consider again the Lyapunov function

$$W_n := \frac{1 - p_n(\alpha)}{p_n(\alpha)}$$

Notice that in view of assumption (3.102), relations (3.72) and (3.75) are fulfilled, i.e.,

$$\widetilde{c}_n(i) - f(x(i)) \overset{a.s.}{=} o_\omega(\frac{1}{n^{1/2-\tau-\varepsilon}}), \ \mathbf{E}\left\{\widetilde{\xi}_n \middle| \, u_n = u(i)\right\} =$$

$$= \Delta(i) + o(\frac{1}{n^{1-2\tau}}) \tag{3.105}$$

Taking into account the Varshavskii-Vorontsova scheme (3.99) and (3.105), it follows:

$$\mathbf{E}\left\{W_{n+1}/\mathcal{F}_n\right\} \overset{a.s.}{=} \sum_{i=1}^{N} \mathbf{E}\left\{W_{n+1}/\mathcal{F}_{n-1} \bigwedge u_n = u(i)\right\} p_n(i) =$$

$$= \sum_{i \neq \alpha}^{N} \mathbf{E}\left\{\frac{1}{p_n(\alpha)\left[1 - \gamma_n p_n(i))(1 - 2\widetilde{\xi}_n)\right]} - 1/\mathcal{F}_{n-1} \bigwedge u_n = u(i)\right\} p_n(i) +$$

$$+\mathbf{E}\left\{\frac{1}{p_n(\alpha) + \gamma_n p_n(\alpha))(1 - 2\widetilde{\xi}_n)\,(1 - p_n(\alpha))} - 1/\mathcal{F}_{n-1} \bigwedge u_n = u(i)\right\} \times$$

$$\times p_n(\alpha) =$$

$$= \sum_{i \neq \alpha}^{N} \mathbf{E}\left\{\frac{1}{p_n(\alpha)}\left[1 + \gamma_n p_n(i))(1 - 2\widetilde{\xi}_n) + O(\gamma_n^2)\right] - 1 \,/\right.$$

$$\left. / \ \mathcal{F}_{n-1} \bigwedge u_n = u(i)\right\} \times p_n(i) -$$

$$-\mathbf{E}\left\{\gamma_n(1 - 2\widetilde{\xi}_n)\,(1 - p_n(\alpha)) + O(\gamma_n^2)/\mathcal{F}_{n-1} \bigwedge u_n = u(i)\right\} =$$

$$= \sum_{i \neq \alpha}^{N} \left(\frac{1}{p_n(\alpha)}\left[1 + \gamma_n p_n(i))\left[1 - 2\Delta(i)\right] + o(\frac{\gamma_n}{n^{1-2\tau}}) + O(\gamma_n^2)\right] - 1\right) p_n(i) -$$

$$-\gamma_n\left[1 - 2\Delta(\alpha)\right](1 - p_n(\alpha)) + o(\frac{\gamma_n}{n^{1-2\tau}}) + O(\gamma_n^2) =$$

$$= W_n\left[1 - p_n(\alpha)(1+\gamma_n)\right] + \frac{\gamma_n}{p_n(\alpha)} \sum_{i \neq \alpha}^{N} p_n^2(i))\left[1 - 2\Delta(i)\right]$$

$$+ o\left(\frac{\gamma_n}{n^{1-2\tau} p_n(\alpha)}\right) + O(\gamma_n^2 p_n^{-1}(\alpha)) = \tag{3.106}$$

for $n \geq n_0(\omega)$, $n_0(\omega) \overset{a.s.}{<} \infty$.

Let us now maximize the second term in (3.106) with respect to the components $p_n(i)$ $(i \neq \alpha)$ under the constraint

$$\sum_{i \neq \alpha}^{N} p_n(i) = 1 - p_n(\alpha) \tag{3.107}$$

To do that, let us introduce the following variables

$$x_i := p_n(i), \quad a_i := \gamma_n\left[1 - 2\Delta(i)\right], \quad i \neq \alpha$$

It follows that

$$\sum_{i \neq \alpha}^{N} p_n^2(i))\left[1 - 2\Delta(i)\right] = \sum_{i \neq \alpha}^{N} a_i x_i^2 := F(x) \tag{3.108}$$

To maximize the function $F(x)$ under the constraint (3.107), let us introduce the following Lagrange function:

$$L(x, \lambda) := F(x) - \lambda\left[\sum_{i \neq \alpha}^{N} x_i - (1 - x_\alpha)\right]$$

The optimal solution (x^*, λ^*) satisfies the following optimality conditions:

$$\frac{\partial}{\partial x_i} L(x^*, \lambda^*) = 2a_i x_i - \lambda^* = 0 \quad \forall i \neq \alpha$$

$$\frac{\partial}{\partial \lambda} L(x^*. \lambda^*) = \sum_{i \neq \alpha}^{N} x_i^* - (1 - x_\alpha) = 0$$

From these optimality conditions, it can be seen that

$$x_i^* = \frac{\lambda^*}{2a_i}, \quad \lambda^* \sum_{i \neq \alpha}^{N} (2a_i)^{-1} = 1 - x_\alpha, \; \lambda^* = \frac{1 - x_\alpha}{\sum\limits_{i \neq \alpha}^{N} (2a_i)^{-1}}$$

Hence

$$F(x) \leq F(x^*) = \frac{1 - x_\alpha}{\sum\limits_{i \neq \alpha}^{N} a_i^{-1}} (N-1) = b(N-1)(1 - x_\alpha)$$

From this inequality we derive

$$\frac{\gamma_n}{p_n(\alpha)} \sum_{i \neq \alpha}^{N} p_n^2(i)) \left[1 - 2\Delta(i)\right] \leq \frac{\gamma_n b\,(N-1)\,(1 - p_n(\alpha))}{p_n(\alpha)} =$$

$$= -\gamma_n\,|b|\,(N-1)\,W_n \qquad (3.109)$$

where

$$b := \left(\sum_{i \neq \alpha}^{N} \left[1 - 2\Delta(i)\right]^{-1} \right)^{-1} < 0 \qquad (3.110)$$

Substituting (3.109) into (3.106), we obtain

$$\mathbf{E}\left\{W_{n+1}/\mathcal{F}_n\right\} \overset{a.s.}{\leq} W_n \left[1 - \; p_n(\alpha)\,(1+\gamma_n) - \gamma_n\,|b|\,(N-1)\right] +$$

$$+ o\left(\frac{\gamma_n}{n^{1-2\tau} p_n(\alpha)}\right) + O(\gamma_n^2 p_n^{-1}(\alpha)) \leq$$

$$\leq W_n \left[1 - \gamma_n\,|b|\,(N-1)\right] + o\left(\frac{\gamma_n}{n^{1-2\tau}} \prod_{t=1}^{n} (1-\gamma_t)^{-1}\right) + O\left(\gamma_n^2 \prod_{t=1}^{n} (1-\gamma_t)^{-1}\right) \qquad (3.111)$$

Taking into account the conditions (3.100) and (3.102), and in view of Robbins-Siegmund theorem [23], we obtain

$$W_n \underset{n \to \infty}{\overset{a.s.}{\to}} 0, \quad p_n(\alpha) \underset{n \to \infty}{\overset{a.s.}{\to}} 1$$

Theorem is proved. ■

Corollary 9. *For example, condition (3.101) will be satisfied for the class of environments for which*

$$\Delta(i) > \frac{1}{2} \quad \forall i \neq \alpha$$

This means that the minimal value of the optimized function must be strictly enough different from its values in the other points.

The following corollary gives an example of the class of the correction factor (gain sequence) satisfying condition (3.101).

Corollary 10. *If we select*

$$\gamma_n = \frac{\gamma}{n+a}, \; \gamma < a \in (0,1]$$

Then, the rate of optimization can be estimated as follows

$$n^\nu W_n \underset{n\to\infty}{\overset{a.s.}{\to}} 0$$

where the order of optimization satisfies the following conditions

$$0 < \nu < \min\left\{1 - 3\gamma, \, |b|^{-1}(N-1)^{-1}\right\} := \nu(\gamma)$$

and the best order of the convergence is equal to

$$\nu(\gamma^*) = \nu^* = |b|^{-1}(N-1)^{-1}$$

and is reached for

$$\gamma = \gamma^* = \frac{1}{3}\left(1 - |b|^{-1}(N-1)^{-1}\right) < a$$

Proof.

follows directly from Lemma A.14 from [14] for

$$\begin{aligned} u_n \; &: = W_n, \; \alpha_n := \gamma_n = \frac{\gamma}{n+a}, \nu_n := n^\nu \\ \beta_n \; &: = \min\left\{o\left(\frac{\gamma_n}{n^{1-2\tau}}\prod_{t=1}^{n}(1-\gamma_t)^{-1}\right); O\left(\gamma_n^2\prod_{t=1}^{n}(1-\gamma_t)^{-1}\right)\right\} = \\ &= \min\left\{o\left(\frac{1}{n^{2-3\gamma}}\right); O\left(\frac{1}{n^{2-\gamma}}\right)\right\} = o\left(\frac{1}{n^{2-3\gamma}}\right) \end{aligned}$$

We obtain

$$\nu < 1 - 3\gamma, \; \nu\,|b|\,(N-1) < 1$$

and hence

$$0 < \nu < \min\left\{1 - 3\gamma, \, |b|^{-1}(N-1)^{-1}\right\} := \nu(\gamma) \le \max_{\gamma < a \in (0,1]} \nu(\gamma) := \nu^*$$

The result of this corollary follows from the last formula by noticing that the maximum of $\nu(\gamma)$ on γ is reached when

$$1 - 3\gamma = |b|^{-1}(N-1)^{-1}$$

Corollary is proved. ■

The point of this theorem is that, a learning automaton using the Varshavskii-Vorontsova reinforcement scheme with the normalization procedure described above, has an asymptotically optimal behaviour, i.e., this reinforcement scheme solves the stochastic optimization problem on finite sets if condition (3.101) is fulfilled.

It is interesting to note that the previous theorems do not require any conditions (convexity, unimodality, differentiability, etc.) on the function to be optimized, as do most optimization techniques.

In this part we have shown that learning automata with fixed number of actions operating in an S-model environment and using a normalization procedure, can successfully solve the stochastic optimization problem on finite sets and, as a result, to solve with an $\varepsilon-$accuracy the stochastic optimization problems related to nonconvex and nondifferentiable functions given on any compact sets. This approach has been already used for the adaptive selection of the optimal order of linear regression models [24].The Bush-Mosteller reinforcement scheme with normalized automaton input has been used to adjust the probability distribution. It has been shown that this approach, while requiring from the user no special skills in stochastic processes, can be a practical tool that open the way to new applications in process modelling, control and optimization.

Another way to solve these stochastic optimization problems is to consider the behaviour of S-model learning automaton with changing number of actions. This approach will be briefly described in the next section.

3.7 Learning Stochastic Automaton with Changing number of Actions

An extensive literature has been dedicated to the behaviour of learning automata with a fixed action set [10]. The concept of the behaviour of automata where the number of the actions available at each time is time-varying has been studied by Thathachar and coauthors [15]. Convergence and convergence rate results have been stated by Poznyak and Najim [17] for learning automata when the action set is time-varying and the input received is continuous.

The set of all actions of the automaton will be denoted by $V = \{u(1), u(2), u(3), ..., u(N)\}$, $2 \leq N < \infty$. This set will be partitioned into W subsets $\left(W = \sum\limits_{k=2}^{N} C_N^k = 2^N - N - 1 \right)$, where $V(j)$ represents the jth action subset. The index j is assigned by ordering these subsets in a lexicographic manner [15] beginning with double-action subset, etc. and ending with the complete set of actions V.

The probability distribution Q_n, defined over all possible action subsets is

$$Q_n = \{q_n(1),\ q_n(2), ..., q_n(W)\}^T$$

where

$q_n(i) = prob\,[V_n = V(i)]$, V_n is the subset selected at time n.

$\sum_{i=1}^{W} q_n(i) = 1. \quad \forall n.$

This probability distribution is assumed to be a priori known.

The learning automaton which has been considered in this study, operates as follows:

Let V_n be the action subset selected at time n. The probability distribution p_n is adjusted according to the following stages [17]

1) Calculate

$$K_n(j) = \sum_{i:u(i)\,\in\,V(j)} p_n(i) > 0$$

$$V_n = V(j)$$

2) Scale the probabilities of the actions in the selected subset

$$p_n^*(i/j) = p_n(i)/K_n(j) \quad for\ all\ i,\ such\ that\ u(i) \in V_n$$

3) Select randomly an action u_i from the subset V_n, according to the scaled probability distribution p_n^*.

4) Use the Bush-Mosteller scheme to adapt p_n^*. The probabilities $p_n(i)$ such that $u_n \notin V_n$, remain unchanged.

5) Rescale the action probabilities of the actions of the selected subset.

For all i, such that $u(i) \in V_n$, do

$$p_n(i) = p_n^*(i/j)K_n(j)$$

So, the probability distribution $P(n)$ is adjusted as follows [17]:

$$p_{n+1}^*(i/j) = p_n^*(i/j) + \gamma_n \left\{ e^{N(j)}(u_n) - p_n^*(i/j) + \right.$$

$$\left. +\xi_n \frac{\left[e^{N(j)} - N(j)e^{N(j)}(u_n)\right]}{N(j)-1} \right\}$$

$$p_{n+1}^*(i/j) = p_n^*(i/j), \quad i \neq j$$

where $N(j)$ denotes the number of actions of the subset V_j,

$2 \le N(j) \le N$.

Under some assumptions, the asymptotic properties of this learning system have been stated in [16]. It has been shown among other that

1. the learning automaton generates asymptotically an optimal pure strategy.

$$e_\alpha^N = (\underbrace{\underbrace{0,..,0,1}_{\alpha},0,...0}_{N})$$

i.e.,

$$p_n(\alpha) \underset{n\to\infty}{\overset{a.s.}{\longrightarrow}} 1, \quad f(x_\alpha) = \min_j f(x_j) < \min_{j\neq\alpha} f(x_j)$$

2. for $N = 3$, the "orders of learning" for automata with changing number of actions is equal to the "orders of learning" of automata with constant number of actions, i.e.,

$$\rho^*_{chang} = \rho^*_{const}$$

3. the desired accuracy ε is achieved by a single learning automaton which has a number of control actions (outputs) N not less than $\frac{D}{\varepsilon} \max_i L_i$.

Hence, the change of action set from instant to instant slows down the learning process. Nevertheless, the behaviour of automata with changing number of actions is useful for solving some engineering problems (optimization, etc.).

We have presented iterative methods for solving stochastic unconstrained optimization problems using learning automata with fixed and changing number of actions and several commonly used reinforcement schemes. Results concerning the convergence and the estimation of the convergence rate have been stated.

We shall illustrate by means of a numerical example how the optimization techniques based on learning automata with fixed and time-varying number of actions may be used to optimize a multimodal function.

3.8 Simulation results

The objective of this section is to give a glimpse of the power of the optimization approaches presented in the previous sections. The learning automata with continuous input (S-model environment) and with or without variable subsets of actions (changing number of actions) described in the preceding Sections were used to optimize the following multimodal function

$$f(x) = 0.3 + \frac{1}{10}\left(\sum_{i=2}^{4} \cos(i-1)\pi\frac{x}{100} + \sum_{i=2}^{4} \sin(i-1)\pi\frac{x}{100}\right)$$
$$x \in [108.5, 198.5]$$

which is depicted in Figure 3.3.

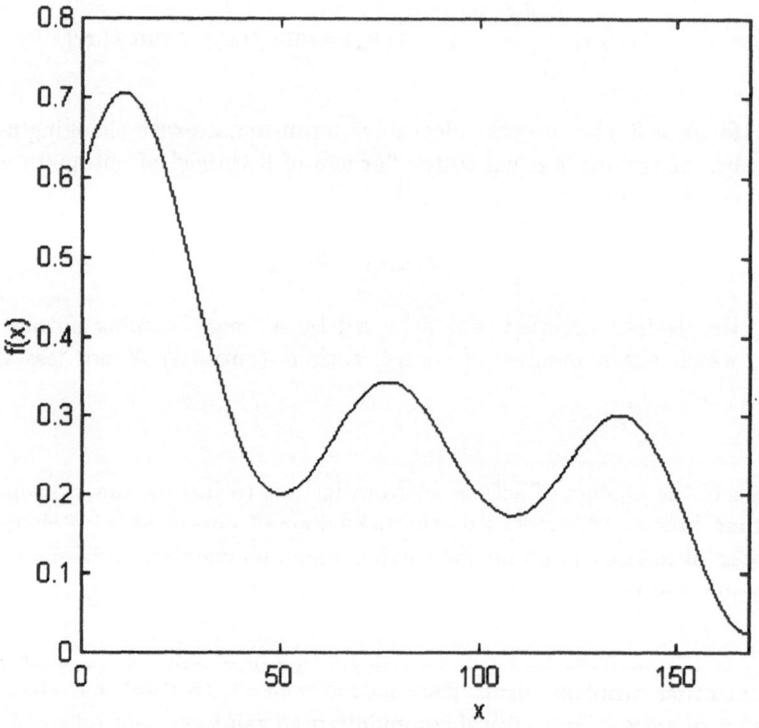

FIGURE 3.3. Multimodal function.

The next subsection deals with the use of learning automata with fixed number of states for multimodal functions optimization.

3.8.1 Learning automaton with fixed number of actions

A learning automaton is used to optimize the multimodal function shown in Figure 3.3. The environment response corresponds to the realization of the function to be optimized. The action selected by the automaton play the role of the environment input, the argument x. At each time n (iteration), the optimization algorithm, based on a learning automata with fixed number of actions performs the following steps:

Step 1. Choice of an action $u(i)$ on the basis of the probability distribution p_n. The technique used by the algorithm to select one subset $u(i)$ among N actions is based on the generation of a normally distributed random variable (any specific machine routine; e.g., RANDU, can be used to carry out a normally distributed random variable).

Step 2. Compute the S-model environment response.

Step 3. Use of this response to adjust the probability distribution according to adopted reinforcement scheme.

Step 4. Return to step 1.

The interval $[108.5, 198.5]$ was partitioned uniformly into 10 values. No prior information was utilized in the initialization of the probabilities, i.e.,

$$p_0 = \left[\frac{1}{N}, \ldots, \frac{1}{N}\right]^T$$

The learning automaton started with a purely uniform distribution for each action. The correction factor γ was chosen to be equal to 0.35.

The evolution of the probabilities (for Bush-Mosteller (B-M), Shapiro-Narendra (S-N) and Varshavskii-Vorontsova (V-V) reinforcement schemes) associated with the optimal action (solution) are depicted in Figure 3.4.

These probabilities tend to one. Figure 3.5 represents the evolution of the loss functions associated with the previous reinforcement schemes.

In practice, it is impossible to obtain perfect measurements. To make the simulations more realistic and to test the performance of the optimization algorithm in the case of noisy observations, a disturbance of zero mean and finite variance equal to 0.02 has been added to the realizations (observations) of the optimized function $f(x)$. The probabilities associated with the optimal solution ($x^* = 168.5$) and related to different reinforcement schemes are depicted in Figure 3.6.

Figure 3.7 shows the evolution of the loss functions associated respectively with Bush-Mosteller (B-M), Shapiro-Narendra (S-N) and Varshavskii-Vorontsova (V-V) reinforcement schemes.

The results are similar when disturbance is not present in the system. The simulation results show that the reinforcement schemes (Bush-Mosteller, Shapiro-Narendra and Varshavskii-Vorontsova) converge to the optimal action. This behaviour was expected from the theoretical results stated in this chapter. The above results can be explained by the adaptive structure of

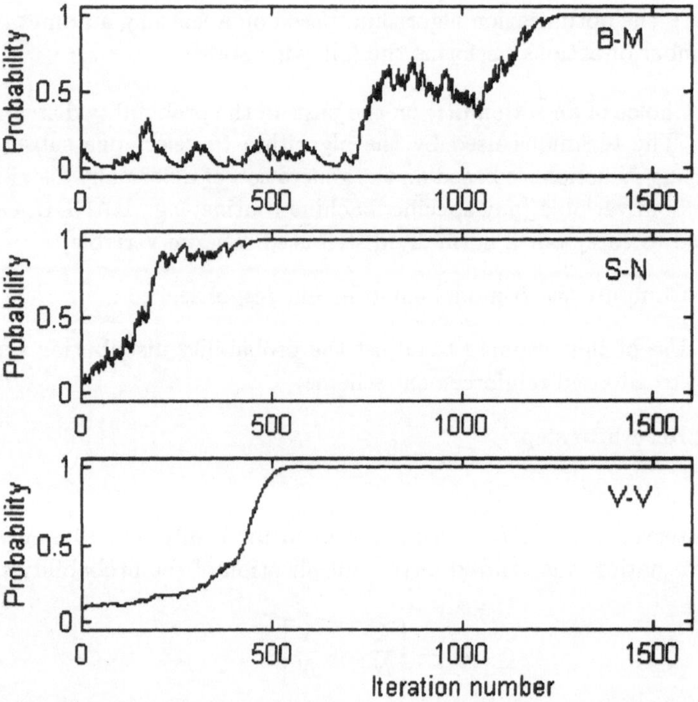

FIGURE 3.4. Evolution of the probabilities associated with the optimal action.

the algorithm and by the fact that a each time (iteration), an action is randomly selected on the basis of the environment response and on the probability distribution p_n.

The next subsection presents some numerical simulations, from which we can verify the viability of the design and analysis concerning the use of learning automata with changing number of actions and which are given in the second part of this chapter.

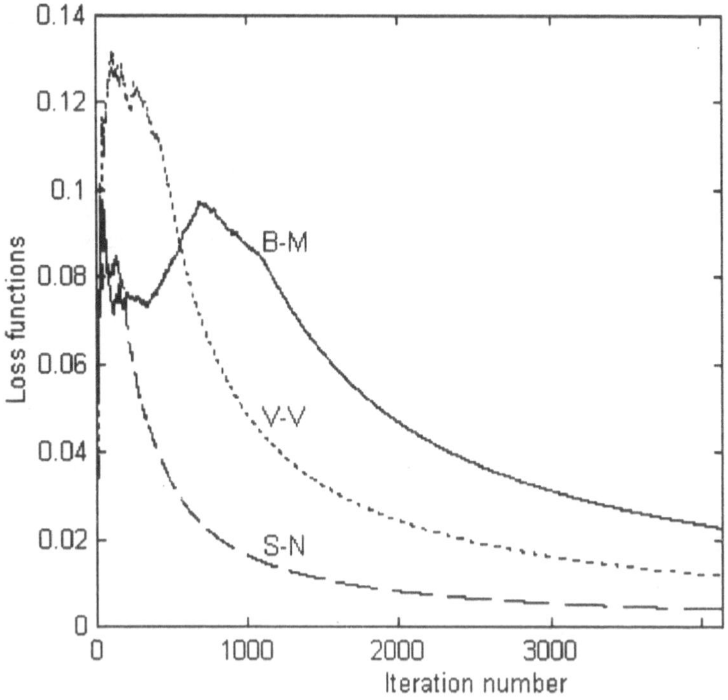

FIGURE 3.5. Evolution of the loss functions.

3.8.2 Learning automaton with changing number of actions

For every time, the optimization algorithm based on learning automata with changing number of actions performs the following steps:

Step1. Choice of a subset $V(j)$ on the basis of the probability Q_n. The technique used by the algorithm to select one subset $V(j)$ among W subsets is based on the generation of a normally distributed random variable (any specific machine routine; e.g., RANDU, can be used to carry out a normally distributed random variable).

Step 2. Choice of an action $u(i)$on the basis of the probability distribution p_n^*. Use of the some procedure as in step 1.

Step 3. Compute the S-model environment response.

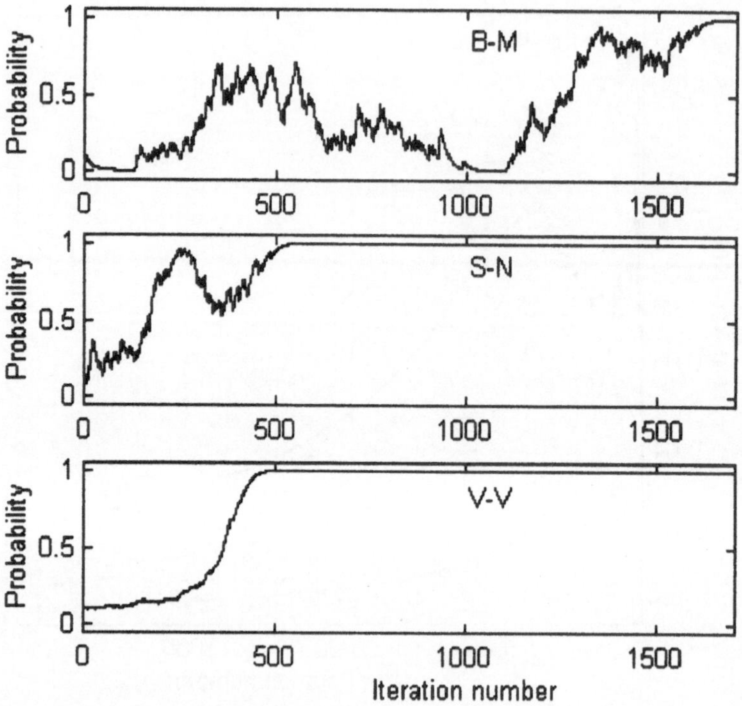

FIGURE 3.6. Evolution of the probabilities associated with the optimal action.

Step 4. Use of this response to adjust the probability distribution according to the given reinforcement scheme.

Step 5. Return to step 1.

The interval [108.5, 198.5] was partitioned uniformly into 10 values (it is up to designer to specify the number of actions of the automaton). The action number can be selected according to the desired accuracy. No prior information was utilized in the initialization of the probabilities. The learning automaton started with a purely uniform distribution for each action, i.e.,

$$p_0 = \left[\frac{1}{N}, ..., \frac{1}{N}\right]^T$$

and

$$Q = \left[\frac{1}{N}, ..., \frac{1}{N}\right]^T$$

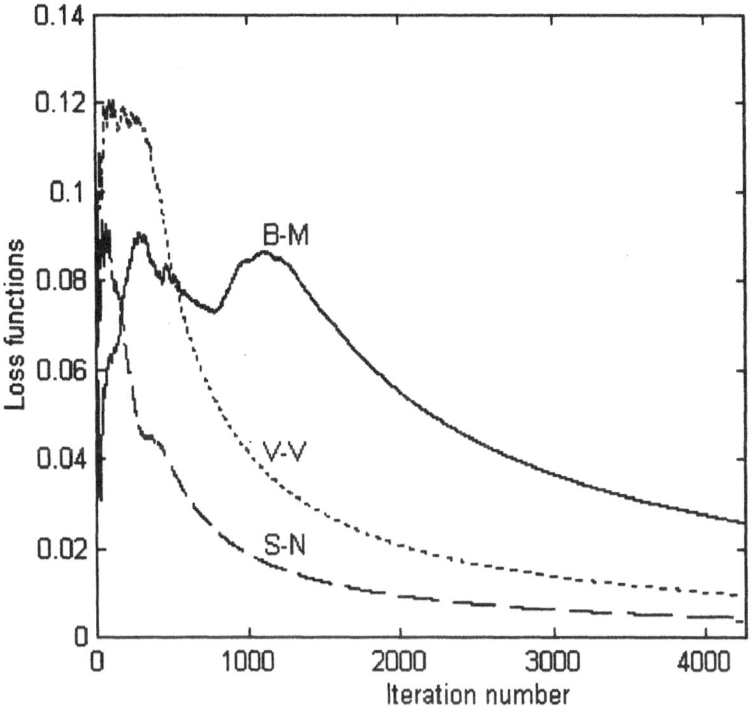

FIGURE 3.7. Evolution of the loss functions.

The theory of learning systems assures that very little a priori information is known about the random environment (the function to be optimized). Nevertheless, the eventually existing information can be introduced through the initial probability distribution or in the design of the environment response (normalization , etc.). The value of the correction factor γ was determined empirically. The value $\gamma = 0.35$ was chosen after some experimentation. Several experiments have been carried out. Figure 3.8 shows the evolution of the probability associated with the optimal state selected by the automaton. This optimal state corresponds to the value of x ($x^* = 168.5$) which minimizes the function $f(x)$ in the interval $[108.5, 198.5]$. The evolution of the loss function is depicted in Figure 3.9.

To test the performance of the optimization algorithm in the case of noisy observations (although the data from real systems are passed through a linear filter to remove aspects of data related to disturbances). A zero mean disturbance was added to the function $f(x)$. The dispersion of this disturbance is equal to 0.02.

Figures 3.10 and 3.11 illustrate the performance of this optimization algorithm. Figure 3.10 represents the evolution of the probability associated

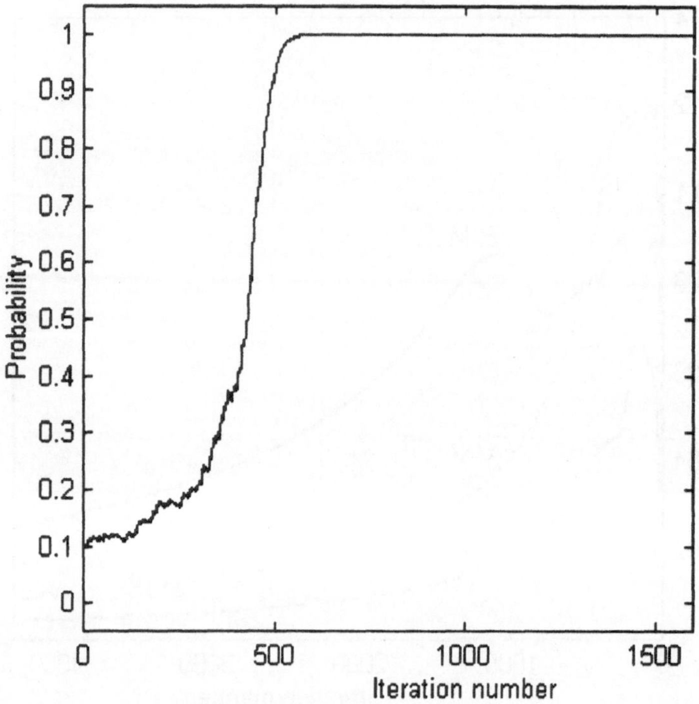

FIGURE 3.8. Evolution of the probability associated with the optimal action.

with optimal action (solution). After approximately 300 iterations, this probability tends to one. The loss function which tends to zero is depicted in Figure 3.11. As a result of the effect of the noise (perturbation) the time needed for obtaining the optimal solution increases.

The decay of the loss function is clearly revealed in Figure 3.11. The nature of convergence of the optimization algorithm is clearly reflected by Figures 3.4-3.11. However, not only is the convergence of optimization algorithm important but the convergence speed is also essential. It depends on the number of operations performed by the algorithm during an iteration as well as the number of iterations needed for convergence. The studied reinforcement schemes behave as expected. They need fewer programming steps and less storage memory.

In summary, to verify the results obtained by theoretical analysis a number of simulations have been carried out with a multimodal function. Notice that the suitable choice of the number of actions N is an important issue and depends on the desired accuracy. In order to overcome the high dimensionality of the number of actions a hierarchical structure of learning

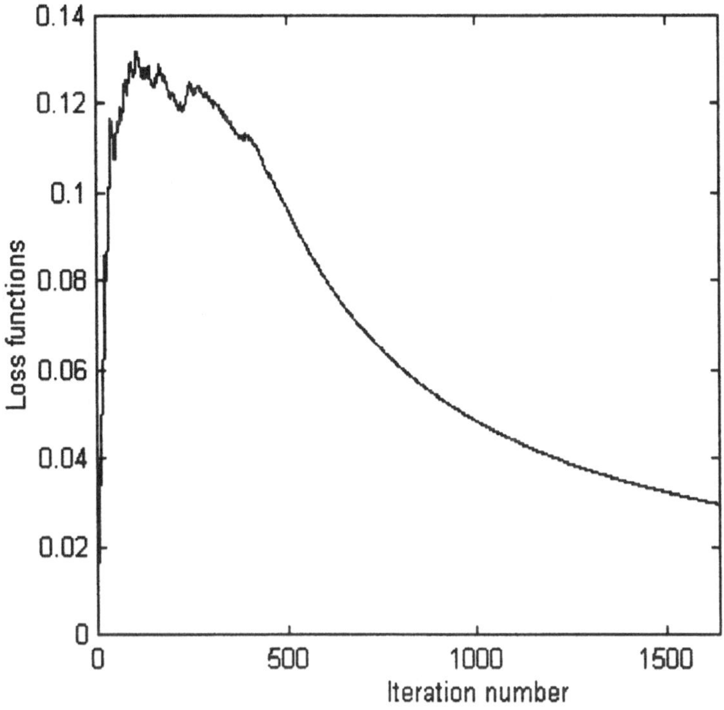

FIGURE 3.9. Evolution of the loss function.

automata can be considered.

In this section we have presented the practical aspects associated with the use of learning automata with fixed and time-varying number of actions for solving unconstrained stochastic optimization problems. The presented simulation results illustrate the performance of this optimization approach, and show that complex structures as multimodal functions, etc. can be learned by learning automata. Other applications were carried out on the basis of the algorithms presented in this chapter. They concern processes control and optimization [25], neural networks synthesis [6] [26], fuzzy logic processor training [27] and model order selection [24]. In [25] A learning automaton with continuous input has been used to solve an optimization problem related to the static control of a continuous stirred tank fermenter. In this study, the discrete values of the manipulated variable (dilution rate) were associated with the actions of the automaton. Neural networks synthesis can be stated as an optimization problem involving numerous local optima and presenting a nonlinear character. The experimental results presented in [6] and [26] show the performance of learning automata when used

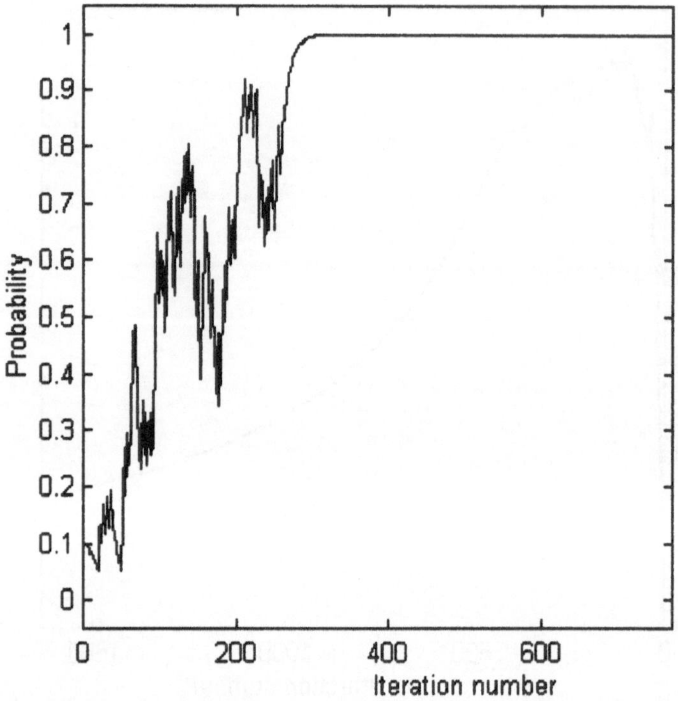

FIGURE 3.10. Evolution of the probability associated with the optimal action.

as optimzers. In [27] a novel method to train the rule base of a fuzzy logic processor is presented. This method is based on learning automata. If compared to more traditional gradient based algorithms, the main advantages are that the gradient need not to be computed (the calculation of the gradients is the most time-consuming task when training logic processors with gradient methods), and that the search for global minimum is not fooled by local minima. An adaptive selection of the optimal order of linear regression models using variable-structure stochastic learning automaton has been presented in [24]. The Akaike criterion has been derived and evaluated for stationary and nonstationary cases. The Bush-Mosteller reinforcement scheme with normalized automaton input has been used to adjust the probability distribution. Simulation results have been carried out to illustrate the feasibility and the performance of this model order selection approach. It has been shown that this approach, while requiring from the user no special skills in stochastic processes, can be a practical tool that open the way to new applications in process modelling, control and optimization.

The authors would like thank Dr. Enso Ikonen for his assistance in carrying out these simulation results.

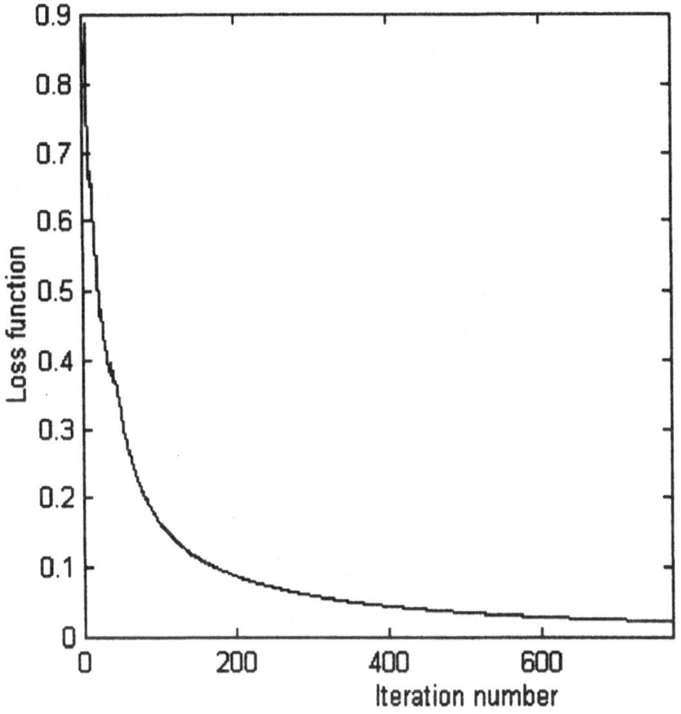

FIGURE 3.11. Evolution of the loss function.

The reinforcement schemes implemented in this chapter give very satisfactory results, and the analytically predictable behaviour is one of their important advantages over heuristic approaches like genetic algorithms and simulated annealing.

3.9 Conclusion

The aim of this chapter is the development of a useful framework for computational method for stochastic unconstrained optimization problems on finite sets using variable-stochastic automata, which incorporate the advantages and flexibility of some of the more recent developments in learning systems. Learning stochastic automata characterized by continuous input and fixed or changing number of actions have been considered. The analysis of these stochastic automata has been given in a concise and unified manner, and some properties involving the convergence and the convergence rate have been stated. A comparison between the performance of a learning automata with and without changing number of actions has been done. Several simulation results have been carried out to show the performance of learning automata when they are used for multimodal functions optimization. In the next chapter we turn to the development of optimization algorithms on the basis of learning automata, for solving stochastic constrained optimization problems on finite sets.

References

[1] Bi-Chong W, Luus R 1978 Reliability of optimization procedures for obtaining global optimum. *AIChE Journal* 24:619-626

[2] Dorea C C Y 1990 Stopping rules for a random optimization method. *SIAM J., Control and Optimization* 28:841-850

[3] Sorensen D C 1982 Newton's method with a model trust region modification. *SIAM Journal Numer. Anal.* 19:409-426

[4] Dolan W B, Cummings P T, Le Van M D 1989 Process optimization via simulated annealing: application to network design. *AIChE Journal* 35:725-736

[5] McMurtry G J, Fu K S 1966 A variable structure automaton used as a multimodal searching technique. *IEEE Trans. on Automatic Control* 11:379-387

[6] Kurz W, Najim K 1992 Synthese neuronaler netze anhand strukturierter stochastischer automaten. *Nachrichten Neuronale Netze Journal* 2:2-6

[7] Kushner H J 1972 Stochastic approximation algorithms for the local optimization of functions with nonunique stationary points. *IEEE Trans. on Automatic Control* 17:646-654

[8] Poznyak A S, Najim K, Chtourou M 1993 Use of recursive stochastic algorithm for neural networks synthesis. *Applied Mathematical Modelling* 17:444-448

[9] Najim K, Chtourou M 1994 Neural networks synthesis based on stochastic approximation algorithm. *International Journal of Systems Science* 25:1219-1222

[10] Narendra K S, Thathachar M A L 1989 *Learning Automata an Introduction.* Prentice-Hall, Englewood Cliffs

[11] Baba N 1984 *New Topics in Learning Automata Theory and Applications.* Springer-Verlag, Berlin

[12] Lakshmivarahan S 1981 *Learning Algorithms Theory and Applications.* Springer-Verlag, Berlin

[13] Najim K, Oppenheim G 1991 Learning systems: theory and application. *IEE Proceedings~E* 138:183-192

[14] Najim K, Poznyak A S 1994 *Learning Automata: Theory and Applications.* Pergamon Press, Oxford

[15] Thathachar M A L, Harita B R 1987 Learning automata with changing number of actions. *IEEE Trans. Syst. Man, and Cybern.* 17:1095-1100

[16] Najim K, Poznyak A S 1996 Multimodal searching technique based on learning automata with continuous input and changing number of actions. *IEEE Trans. Syst. Man, and Cybern.* 26:666-673

[17] Poznyak A S, Najim K 1997 Learning automata with continuous input and changing number of actions. *to appear in International Journal of Systems Science.*

[18] Bush R R, Mosteller F 1958 *Stochastic Models for Learning.* John Wiley & Sons, New York

[19] Shapiro I J, Narendra K S 1969 Use of stochastic automata for parameter self optimization with multimodal performance criteria. *IEEE Trans. Syst. Man, and Cybern.* 5:352-361

[20] Varshavskii V I, Vorontsova I P 1963 On the behavior of stochastic automata with variable structure. *Automation and Remote Control* 24:327-333

[21] Ash B B 1972 *Real Analysis and Probability.* Academic Press, New York

[22] Doob J L 1953 *Stochastic Processes.* John Wiley & Sons, New York

[23] Robbins H, Siegmund D 1971 A convergence theorem for nonnegative almost supermartingales and some applications. In Rustagi J S (ed) 1971 *Optimizing Methods in Statistics.* Academic Press, New York

[24] Poznyak A S, Najim K, Ikonen E 1996 Adaptive selection of the optimal order of linear regression models using learning automata. *Int. J. of Systems Science* 27:151-159

[25] Najim K, Mészaros A, Rusnak A 1997 A stochastic optimization algorithm based on learning automata. *Journal a*

[26] Najim K, Chtourou M, Thibault J 1992 Neural network synthesis using learning automata. *Journal of Systems Engineering* 2:192-197

[27] Ikonen E, Najim K 1997 Use of learning automata in distributed fuzzy logic processor training. *IEE Proceedings~E*

4

Constrained Optimization Problems

4.1 Introduction

In the previous chapter, we discussed how to solve stochastic unconstrained optimization problems using learning automata. The goal of this chapter is to solve stochastic constrained optimization problems using different approaches: Lagrange multipliers and penalty functions.

Optimization is a large field of numerical research. Engineers (electrical, chemical, mechanical, etc.) and economist are often confronted with constrained optimization where there are a priori limitations (constraints) on the allowed values of independent variables or some functions of these variables.

There is now a strong and growing interest within the engineering community in the use of efficient optimization techniques for several purposes. Optimization techniques are commonly used as a framework for the formulation and solution of design problems [1]-[2]. For example, in the context of control, the objective of an on-line optimization scheme is to track the real process optimum as it changes with time. This must be achieved while allowing for disturbances to the process, ensuring process constraints are not violated. In recent years, learning automata [3]-[4] have attracted the attention of scientists and technologists from a number of disciplines, and have been applied to a wide variety of practical problems in which a priori information is incomplete. In fact, observations measured from natural phenomena posses an inherent probabilistic structure. We hasten to note that in stochastic control theory, it is assumed that the probability distribution are known. Unfortunately it is the exception rather than the rule that statistic characteristics of the considered processes are a priori known. In real situations, modelling always involves some element of approximation since all real systems are, to some extent, nonlinear, time-varying and disturbed [5].

Descent algorithms such as steepest descent, conjugate gradient, quasi Newton methods, methods of feasible directions, projected gradient method with trust region, etc. are commonly used to solve optimization problems with or without constraints [6]-[7]-[8]-[9]-[10]-[11]-[12]-[13]. They are based on direct gradient measurements. When only noisy measurements are available, algorithms based on gradient approximation from measurements, stochastic approximation algorithms and random search techniques [14] are appropriate for solving stochastic optimization problems. The main advantage of random search over other direct search techniques is its general applicability, i.e., there are almost no conditions concerning the function to be optimized (continuity, unimodality, convexity, etc.). Methods based on learning systems [3]-[15]-[16] belong to this class of random search techniques.

The aim of this chapter is the development of a useful framework for computational method for stochastic constrained optimization problems on finite sets using variable-stochastic automata, which incorporate the advantages and flexibility of some of the more recent developments in learning systems [4].

In recent years, learning systems [4]-[17]-[18] have been fairly exhaustively studied. This interest is essentially due to the fact that learning systems provide interesting methods for solving complex nonlinear problems characterized by a high lèvel of uncertainty. Learning is defined as any relatively permanent change in behaviour resulting from past experience, and a learning system is characterized by its ability to improve its behaviour with time, in some sense tending towards the ultimate goal. They have been used for solving unconstrained optimization problems [4].

This stochastic constrained optimization problem is equivalent to the stochastic linear programming problem which is formulated and solved as the behaviour of a variable-stochastic automaton in a multi-teacher environment [4]-[19]. A reinforcement scheme derived from the Bush-Mosteller scheme [15] is used to adapt the probabilities associated with the actions of the automaton. The learning automaton operates in a multi-teacher environment. The environment response which is constructed from the available data (cost function and constraints realizations) has been normalized and used as the automaton input.

In this chapter we shall be concerned with stochastic constrained optimization problems on finite sets using two approaches: Lagrange multipliers and penalty functions. The first part of this chapter is dedicated to the Lagrange multipliers approach. It is shown that the considered optimization problem is equivalent to the stochastic linear programming problem given on a multidimensional simplex. The Lagrange function associated with the later is not strictly convex and, as a result of this fact, any attempts to apply directly the gradient technique for finding its saddle-point doomed to failure. To avoid this problem we have introduced a regularization term in the corresponding Lagrange function as in [20]. The Lipschitzian property for the saddle-point of the regularized Lagrange function with respect to the regularization parameter is proved.

The second part of this chapter presents an alternative approach for solving the same stochastic constrained optimization problem. It is based on learning automata and the penalty function approach [12]-[21]-[22]-[23]-

[24]-[25]-[26]-[27]. The environment response is processed using a new normalization procedure. The properties of the optimal solution as well as the basic convergence characteristics of the optimization algorithm are stated.

Our main theoretical results (convergence analysis) are stated using martingales theory and Lyapunov approach. In fact, martingales arise naturally whenever one needs to consider conditional expectation with respect to increasing information patterns [5]-[28].

4.2 Constrained optimization problem

4.2.1 Problem formulation

Let us consider the following sequence of random vectors

$$\zeta_n^j(u,\omega),\ u \in U := \{u(1),...,u(N)\}\ ;\ \omega \in \Omega;\ n = 1,2,...;j = 0,1,...,m$$

which are given on a probability space $(\Omega, \mathcal{F}, \mathbf{P})$ where

- u is a discrete variable defined on the set U
- ω is a random variable defined on the set of elementary events Ω.

Let us define the Borel functions Φ_n^j $(j = 0,1,...,m)$ as follows

$$\Phi_n^j := \frac{1}{n}\sum_{t=1}^{n} \zeta_t^j(u_t,\omega) \tag{4.1}$$

where the sequence $\{u_t\}_{t=1,...,n}$ of random variables is also defined on $(\Omega, \mathcal{F}, \mathbf{P})$ and represents the optimization sequence *"optimization strategy"* which depends on the available information, i.e.,

$$u_n = u_n\left(\zeta_t^j, u_s, \omega | j = 0,1,...,m\right);\ t = 1,...,n;\ s = 1,...,n-1 \tag{4.2}$$

Let us consider the following assumptions:

- **(H1)** The conditional mathematical expectation of the random variables $\zeta_n^j(u_n,\omega)$ for any fixed random variable $u_n = u(i)$ is stationary, i.e.,

$$\mathbf{E}\left\{\zeta_n^j(u_n,\omega)|u_n = u(i) \wedge \mathcal{F}_{n-1}\right\} \overset{a.s.}{=} v_i^j\ (j = 0,1,...,m; i = 1,...,N)$$

where

$$\mathcal{F}_{n-1} := \sigma(\zeta_t^j, u_s, \omega | j = 0,1,...,m; t = 1,...,n-1; s = 1,...,n-1)$$

is the $\sigma-$algebra generated by the corresponding data.

- **(H2)** For any fixed random variable $u_n = u(i)$, the absolute value of the random variables $\zeta_n^j(u_n, \omega)$ are uniformly bounded with probability one, i.e.,.

$$\mathop{lim\sup}_{n\to\infty} \left|\zeta_n^j(u_n, \omega)\right| \overset{a.s.}{\leq} \sigma_j^+ < \infty \ (j = 0, ..., N)$$

The main stochastic optimization problem treated in this chapter may now be stated:

Under assumptions (H1) and (H2), find the optimization strategy $\{u_n\}$ which minimizes, with probability 1, the average loss function

$$\limsup_{n\to\infty} \Phi_n^0 \overset{a.s.}{\to} \inf_{\{u_n\}} \tag{4.3}$$

under the constraints

$$\limsup_{n\to\infty} \Phi_n^j \overset{a.s.}{\leq} 0; \quad (j = 1, ..., m) \tag{4.4}$$

From (4.3) and (4.4) it follows that the sequence $\{\zeta_n^0\}$ and the sequences $\{\zeta_n^j\}$ $(j = 1, ..., m)$ correspond respectively to the realizations of the loss function Φ_n^0 and the constraints defined by the functions Φ_n^j $(j = 1, ..., m)$.

Remark. *The "chance constraints" [25]-[29]*

$$\mathbf{P}\left\{\omega : \phi_j(\zeta_n^0, \zeta_n^1, ..., \zeta_n^m) \geq \alpha_j\right\} \leq \beta_j \tag{4.5}$$

belong to the class of constraints (4.4). Indeed, this chance constraint can be written as follows:

$$\mathbf{E}\left\{\zeta_n^j\right\} \leq 0 \tag{4.6}$$

where

$$\zeta_n^j = \chi\left\{\phi_j(\zeta_n^0, \zeta_n^1, ..., \zeta_n^m) \geq \alpha_j\right\} - \beta_j \tag{4.7}$$

and the indicator function $\chi(\cdot)$ is defined as follows

$$\chi(\mathcal{A}) = \begin{cases} 1 \text{ if } \mathcal{A} \text{ is true} \\ 0 \text{ otherwise} \end{cases}$$

Substituting (4.7) into (4.4) leads to

$$\limsup_{n\to\infty} \frac{1}{n} \sum_{t=1}^{n} \chi\left\{\phi_j(\zeta_t^0, \zeta_t^1, ..., \zeta_t^m) \geq \alpha_j\right\} \leq \beta_j \tag{4.8}$$

In the stationary case, i.e., the probability distribution of the vector $(\zeta_t^0, \zeta_t^1, ..., \zeta_t^m)^T$ is stationary, and in view of the strong law of large numbers [28]-[30], it follows that the inequalities (4.8) and (4.5) are equivalent, i.e.,

$$\limsup_{n\to\infty} \frac{1}{n}\sum_{t=1}^{n} \chi\left\{\phi_j(\zeta_t^0,\zeta_t^1,...,\zeta_t^m) \geq \alpha_j\right\}$$

$$= \limsup_{n\to\infty} \frac{1}{n}\sum_{t=1}^{n} \mathbf{E}\left\{\chi\left\{\phi_j(\zeta_t^0,\zeta_t^1,...,\zeta_t^m) \geq \alpha_j\right\}\right\} =$$

$$= \mathbf{P}\left\{\omega : \phi_j(\zeta_n^0,\zeta_n^1,...,\zeta_n^m) \geq \alpha_j\right\} \leq \beta_j$$

The problem (4.3)-(4.4) is asymptotically equivalent to a linear programming problem which will be stated in the next subsection.

4.2.2 EQUIVALENT LINEAR PROGRAMMING PROBLEM

Consider the following stochastic programming problem on a multidimensional simplex:

$$V_0(p) := \sum_{i=1}^{N} v_i^0 p(i) \to \sup_{p\in S} \tag{4.9}$$

$$V_j(p) := \sum_{i=1}^{N} v_i^j p(i) \leq 0, \quad (j = 1, ..., m) \tag{4.10}$$

$$S^N := \left\{ p = (p(1), ..., p(N)) \in R^N \mid p(i) \geq 0, \sum_{i=1}^{N} p(i) = 1 \right\} \tag{4.11}$$

where S^N denotes the simplex in the Euclidean space R^N and p is the probability distribution along with the optimization is done.

The following theorem states the asymptotic equivalence between the basic problem (4.3)-(4.4) and the linear programming problem (4.9)-(4.11).

Theorem 1. *Under assumptions (H1) and (H2):*

1. *The basic stochastic constrained optimization problem (4.3)-(4.4) has a solution if and only if the corresponding linear programming problem (4.9)-(4.11) has a solution too (may be not unique).*
2. *These solutions coincide with probability one, i.e.,*

$$\inf_{\{u_n\}} \Phi_n^0 \overset{a.s.}{=} \inf_{p\in S^N} V_0(p)$$

such that the constraints (4.4) and (4.10) are simultaneously fulfilled.

Proof.

Assume that the basic problem (4.3)-(4.4) has a solution. Then, according to Lemma A.4-1 (appendix A), the sequence $\{u_n\}$ satisfies asymptotically ($n \to \infty$) the constraints (4.9) if and only if the sequence

$$f_n(i) := \frac{1}{n}\sum_{t=1}^{n} \chi(u_t = u(i)) \tag{4.12}$$

satisfy the set of constraints (4.10)-(4.11). Hence,

$$\inf_{\{u_n\}} \Phi_n^0 \overset{a.s.}{\geq} V_0(p^*) \tag{4.13}$$

where $p^* \in S^N$ is the solution of the linear programming problem (4.9)-(4.11). Let us now consider the stationary optimization strategy $\{u_n\}$ for which

$$\mathbf{P}\ \{\omega : u_n = u(i)|\mathcal{F}_{n-1}\} \overset{a.s.}{=} p^*(i)$$

According to the strong law of large numbers for random sequences containing martingale differences (see Lemma A.4-2 in appendix A) we derive

$$\lim_{n\to\infty} \Phi_n^j \overset{a.s.}{=} V_j(p^*)\ (j = 1, ..., m)$$

Theorem is proved. ■

The next section deals with the properties of the Lagrange function and its saddle-point.

4.3 Lagrange multipliers using regularization approach

It is well known [11]-[26]-[27]-[8] that the stochastic linear programming problem discussed above is equivalent to the problem of finding the saddle-point for the following Lagrange function

$$L(p, \lambda) := V_0(p) + \sum_{i=1}^{m} \lambda(j)V_j(p) \tag{4.14}$$

given on the set $S^N \times R_+^m$ where

$$R_+^m := \lambda = \{(\lambda(1), ..., \lambda(m)) : \lambda(i) \geq 0\ i = 1, ..., m\} \tag{4.15}$$

In other words, the initial problem is equivalent to the following minmax problem

$$L(p, \lambda) \to \inf_{p\in S^N} \sup_{\lambda\in R_+^m} \tag{4.16}$$

According to the Lagrange Multipliers Theory [7]-[26], any vector $p^* \in S^N$ is the solution of the linear programming problem (4.9)-(4.11) if and only if there exists a vector $\lambda^* \in R_+^m$ such that for any other vectors $p \in S^N$ and $\lambda \in R_+^m$ the following saddle-point inequalities hold

$$L(p^*, \lambda) \leq L(p^*, \lambda^*) \leq L(p, \lambda^*) \tag{4.17}$$

The Lagrange function $L(p, \lambda)$ is not strictly convex and, as a consequence, any attempt to apply directly the gradient technique for finding its saddle-point doomed to failure.

The following example illustrates this fact.

Example. *Consider the function $L(p, \lambda)$ for which their arguments belong to the single dimensional space ($N = m = 1$), and $V_0(p) = 0$, i.e.,*

$$L(p, \lambda) = \lambda p$$

Applying directly the gradient technique (Arrow-Hurwicz-Uzava algorithm) [31] for finding its saddle-point which is equal to $p^* = \lambda^* = 0$, we obtain

$$\begin{aligned} p_n &= p_{n-1} - \gamma_n \nabla_p L(p_{n-1}, \lambda_{n-1}) = p_{n-1} - \gamma_n \lambda_{n-1} \\ \lambda_n &= \lambda_{n-1} + \gamma_n \nabla_\lambda L(p_{n-1}, \lambda_{n-1}) = \lambda_{n-1} + \gamma_n p_{n-1} \\ \gamma_n &\in R^1, \ \gamma_n \geq 0, \ \sum_{n=1}^{\infty} \gamma_n = \infty \end{aligned} \tag{4.18}$$

It is easy to show that this procedure generates a divergent sequence $\{p_n, \lambda_n\}$ which does not tend to the saddle-point $p^* = \lambda^* = 0$. Indeed,

$$\rho_n := p_n^2 + \lambda_n^2 = \left(1 + \gamma_n^2\right) \rho_{n-1} = \rho_0 \prod_{t=1}^{n} \left(1 + \gamma_t^2\right) \geq \rho_0 > 0$$

One approach for avoiding this problem consists of introducing a regularization term in the corresponding Lagrange function [7]-[20]:

$$L_\delta(p, \lambda) = L(p, \lambda) + \frac{\delta}{2} \left(\|p\|^2 - \|\lambda\|^2 \right) \tag{4.19}$$

For each regularizing parameter $\delta > 0$, this regularized Lagrange function $L_\delta(p, \lambda)$ is strictly convex on p and concave on λ . The next theorem describes the dependence of the saddle-point $(p^*(\delta), \lambda^*(\delta))$ of this regularized function with the regularizing parameter δ and analyses its asymptotic behaviour when $\delta \to 0$.

Theorem 2. *For any sequence $\{\delta_n\}$ such that*

$$0 < \delta_n < 1, \ \delta_n \to 0 \tag{4.20}$$

the sequence $\{(p^*(\delta_n), \lambda^*(\delta_n))\}$ of the saddle-points of the corresponding Lagrange function $L_\delta(p, \lambda)$ (4.14) converges to the saddle point (p^{**}, λ^{**}) of the initial Lagrange function $L(p, \lambda)$, which corresponds to the saddle point (p^*, λ^*) and has the minimal norm, i.e., if there exist many saddle-points (p^*, λ^*) of the initial Lagrange function then,

$$(p^*(\delta_n), \lambda^*(\delta_n)) \rightarrow (p^{**}, \lambda^{**}), \; n \rightarrow \infty$$

where

$$p^{**} := \arg \min_{p^*, \lambda^*} \frac{1}{2}\left(\|p^*\|^2 + \|\lambda^*\|^2\right)$$

and (p^*, λ^*) is the saddle-point of $L(p, \lambda)$.

Proof.

Let us introduce the following notations

$$p_n^* := p^*(\delta_n), \; \lambda_n^* := \lambda^*(\delta_n)$$

Using inequalities (4.17) and definition (4.68) we derive

$$L_\delta(p^*(\delta_n), \lambda^*) = L\left(p^*(\delta_n), \lambda^*\right) + \frac{\delta}{2}\left(\|p^*(\delta_n)\|^2 - \|\lambda^*\|^2\right) \leq$$

$$\leq L\left(p^*, \lambda^*(\delta_n)\right) + \frac{\delta}{2}\left(\|p^*\|^2 - \|\lambda^*(\delta_n)\|^2\right) =$$

$$= L_\delta(p^*, \lambda^*(\delta_n))$$

After dividing by $\delta > 0$, this inequality leads to

$$\|p^*(\delta_n)\|^2 + \|\lambda^*(\delta_n)\|^2 \leq \frac{2}{\delta}\left[L\left(p^*, \lambda^*(\delta_n)\right) - L\left(p^*(\delta_n), \lambda^*\right)\right] +$$

$$+ \|p^*\|^2 + \|\lambda^*\|^2 \leq \|p^*\|^2 + \|\lambda^*\|^2 < \infty$$

From this inequality it follows that the sequence $\{p_n^*, \lambda_n^*\}$ is uniformly bounded on n. Hence, we can select a convergent subsequence $\{p_{n_k}^*, \lambda_{n_k}^*\}_{k\rightarrow\infty}$, i.e., the following limits exist

$$\lim_{k\rightarrow\infty} p_{n_k}^* := \tilde{p}, \; \lim_{k\rightarrow\infty} \lambda_{n_k}^* := \tilde{\lambda}$$

It is clear that this limit point $\left(\tilde{p}, \tilde{\lambda}\right)$ is a saddle-point of the initial Lagrange function $L(p, \lambda)$ (4.14). If it is unique then the theorem is proved.

Now consider the case, when we are concerned with a set of saddle-points. The following inequality

$$0 \leq (p - p_n^*)^T \nabla_p L_\delta(p, \lambda) - (\lambda - \lambda_n^*)^T \nabla_\lambda L_\delta(p, \lambda), \forall\lambda, p \qquad (4.21)$$

holds for any convex function.

Let us suppose that for two different convergent subsequences there exist two different limits, i.e.,

$$\lim_{k\to\infty} p_{n_k}^* := \widetilde{p}, \ \lim_{k\to\infty} \lambda_{n_k}^* := \widetilde{\lambda}$$

$$\lim_{s\to\infty} p_{n_s}^* := \widetilde{\widetilde{p}}, \ \lim_{s\to\infty} \lambda_{n_s}^* := \widetilde{\widetilde{\lambda}}$$

For $p = p^*, \lambda = \lambda^*$ and $n = n_k$ or $n = n_s$, inequality (4.21) and the properties (4.17) of the saddle-point [10] lead to

$$0 \leq (p^* - p_{n_k}^*)^T \left(v^0 + \sum_{j=1}^{m} \lambda_j^* v^j + \delta_{n_k} p^* \right) -$$

$$- (\lambda^* - \lambda_{n_k}^*)^T \left(\left[p^{*T} v^1, ..., p^{*T} v^m \right]^T - \delta_{n_k} \lambda^* \right) =$$

$$= L(p^*, \lambda^*) - L(p_{n_k}^*, \lambda^*) + L(p^*, \lambda_{n_k}^*) - L(p^*, \lambda^*) +$$

$$+ \delta_{n_k} \left[\left(p^* - p_{n_k}^* \right)^T p^* + \left(\lambda^* - \lambda_{n_k}^* \right)^T \lambda^* \right] \leq$$

$$\leq \delta_{n_k} \left[\left(p^* - p_{n_k}^* \right)^T p^* + \left(\lambda^* - \lambda_{n_k}^* \right)^T \lambda^* \right]$$

where

$$v^0 := (v_1^0, ..., v_N^0)^T, \ v^j := (v_1^j, ..., v_N^j)^T \ j = 1, ..., m$$

Dividing both sides of the last inequality by δ_{n_k}, and for $k \to \infty$, we obtain

$$(p^* - \widetilde{p})^T p^* + \left(\lambda^* - \widetilde{\lambda} \right)^T \lambda^* \geq 0$$

for any saddle-points (p^*, λ^*) of the Lagrange function $L(p, \lambda)$. Similarly, for the other subsequence we can write

$$\left(p^* - \widetilde{\widetilde{p}} \right)^T p^* + \left(\lambda^* - \widetilde{\widetilde{\lambda}} \right)^T \lambda^* \geq 0$$

From these last inequalities it follows that the points $(\widetilde{p}, \widetilde{\lambda})$ and $(\widetilde{\widetilde{p}}, \widetilde{\widetilde{\lambda}})$ correspond to the minimum of the following quadratic function

$$\frac{1}{2} \left(\|p^*\|^2 + \|\lambda^*\|^2 \right)$$

defined on the set of all possible saddle-points (p^*, λ^*). This function is strictly convex, hence the minimum is unique. As a consequence, it follows

$$\widetilde{p} = \widetilde{\widetilde{p}}, \quad \widetilde{\lambda} = \widetilde{\widetilde{\lambda}}$$

Theorem is proved. ■

For $\delta > 0$, the regularized Lagrange function is a strictly convex function and, has the following useful property which will be given in the next lemma.

Lemma 1. *For any $p \in S^N$ and for any $\lambda \in R^m$ the following inequality holds:*

$$(p - p^*(\delta))^T \nabla_p L_\delta(p, \lambda) - (\lambda - \lambda^*(\delta))^T \nabla_\lambda L_\delta(p, \lambda) \geq$$

$$\geq \frac{\delta}{2} \left(\|p - p^*(\delta)\|^2 + \|\lambda - \lambda^*(\delta)\|^2 \right)$$

where $p^*(\delta), \lambda^*(\delta)$ is the saddle-point of the regularized Lagrange function (4.68).

Proof.

From (4.68) we obtain

$$(p - p^*(\delta))^T \nabla_p L_\delta(p, \lambda) - (\lambda - \lambda^*(\delta))^T \nabla_\lambda L_\delta(p, \lambda) =$$

$$= L_\delta(p, \lambda^*(\delta)) - L_\delta(p^*(\delta), \lambda) + \frac{\delta}{2} \left(\|p - p^*(\delta)\|^2 + \|\lambda - \lambda^*(\delta)\|^2 \right)$$

The desired result follows immediately from the following inequality

$$L_\delta(p, \lambda^*(\delta)) \geq L_\delta(p^*(\delta), \lambda)$$

which corresponds to the saddle-point condition (4.17). Lemma is proved. ■

The next theorem states the Lipschitzian property for the saddle-point $(p^*(\delta), \lambda^*(\delta))$ of the regularized Lagrange function $L_\delta(p, \lambda)$ with respect to the regularization parameter δ.

Theorem 3. *Under the same conditions as in Theorem 2, there exists a positive constant C such that for any nonnegative parameters δ_t and δ_s the following inequality holds:*

$$\|p^*(\delta_t) - p^*(\delta_s)\| + \|\lambda^*(\delta_t) - \lambda^*(\delta_s)\| \leq C \, |\delta_t - \delta_s|$$

Proof.

Let us consider the following sets

$$Z_0 := \left\{ (p, \lambda) : \sum_{i=1}^{N} p(i) = 1 \right\}$$

$$Z_1(i_1, ..., i_s) := \{(p, \lambda) : p(i_k) = 0,\ k = 1, ..., s\} \cap Z_0$$

$$Z_2(j_1, ..., j_r) := \{(p, \lambda) : \lambda(j_k) = 0,\ k = 1, ..., r\} \cap Z_0$$

$$Z_3(i_1, ..., i_s; j_1, ..., j_r) := Z_1(i_1, ..., i_s) \cap Z_2(j_1, ..., j_r)$$

The total number of these sets is equal to $2^m(2^N - 1)$. They will be denoted by $\mathcal{G}_k$ $\left(k = 1, ..., 2^m(2^N - 1)\right)$. Let us associate with each set $\mathcal{G}_k$ the problem $\mathfrak{P}_{\mathfrak{k}}$ of finding the saddle-point of the regularized Lagrange function $L_\delta(p, \lambda)$. The corresponding solution will be denoted by $(p(\mathfrak{P}_{\mathfrak{k}}),\ \lambda(\mathfrak{P}_{\mathfrak{k}}))$. It is clear that the saddle-point $(p^*(\delta),\ \lambda^*(\delta))$ of the regularized Lagrange function $L_\delta(p, \lambda)$, defined on the set $S^N \times R_+^m$, coincides with the solution $(p(\mathfrak{P}_{\mathfrak{k}}),\ \lambda(\mathfrak{P}_{\mathfrak{k}}))$ of one of the previous problems $\mathfrak{P}_{\mathfrak{k}}$. Notice that each problem $\mathfrak{P}_{\mathfrak{k}}$ is a convex optimization problem with equality constraints and, as a consequence we can use the Lagrange technique for deriving the necessary optimality conditions. The optimality conditions are:

$$\nabla_p \mathcal{L}_\delta(p, \lambda, \mu) = 0, \quad \nabla_\lambda \mathcal{L}_\delta(p, \lambda, \mu) = 0, \quad \nabla_\mu \mathcal{L}_\delta(p, \lambda, \mu) = 0 \qquad (4.22)$$

where

$$\mathcal{L}_\delta(p, \lambda, \mu) := L_\delta(p, \lambda) - \mu_0 \left(\sum_{i=1}^{N} p(i) - 1 \right) - \sum_k \mu_{1k} p(i_k) - \sum_k \mu_{2k} \lambda(j_k)$$

and μ_{1k} and μ_{2k} are Lagrange multipliers.

The linear system of equations (4.22) is equivalent to

$$\begin{aligned} v^0 + \sum_{j=1}^{m} \lambda_j v^j + \delta p - \mu_0 e^N - \sum_k \mu_{1k} e(i_k) &= 0 \\ \left[p^T v^1, ..., p^T v^m\right]^T - \delta\lambda - \sum_k \mu_{2k} e(j_k) &= 0 \end{aligned} \qquad (4.23)$$

$$(p, \lambda) \in \mathcal{G}_k$$

where

$$e^N = (1, ..., 1)^T \in R^N \quad e(j_k) = (\underbrace{0, ..0, 1}_{j_k}, 0, 0, ..., 0)^T$$

From (4.35) it follows that the expression of the optimal solution (saddle-point) can be written in the following parametric form

$$p(i) = \frac{\sum\limits_{s=0}^{m+N} b_{is}^{p}\delta^{s}}{\sum\limits_{s=0}^{m+N} a_{is}^{p}\delta^{s}}, \quad \lambda(j) = \frac{\sum\limits_{s=0}^{m+N} b_{js}^{\lambda}\delta^{s}}{\sum\limits_{s=0}^{m+N} a_{js}^{\lambda}\delta^{s}} \tag{4.24}$$

For any $\delta > 0$, the corresponding Lagrange function is strictly convex and has a unique optimal point. It follows that the denominator of (4.24) is not equal to 0. When δ is small enough the problem to be solved remains the same ($\mathfrak{P}_{\mathfrak{k}}$) as δ decreases. Let us assume that the $r_p^{'}$–first coefficients b_{is}^{p} $\left(s = 0, 1, ..., r_p^{'}\right)$ are equal to zero and, similarly assume that the $r_p^{''}$–first coefficients a_{is}^{p} $\left(s = 0, 1, ..., r_p^{''}\right)$ are also equal to zero. It follows:

$$p(i) = \delta^{r_p} \frac{b_{i,r_p^{'}+1}^{p} + \sum\limits_{s=1}^{m+N-r_p^{'}-1} b_{i,s+r_p^{'}+1}^{p}\delta^{s}}{a_{ir_p^{''}+1}^{p} + \sum\limits_{s=1}^{m+N} a_{i,s+r_p^{''}+1}^{p}\delta^{s}}, \quad r_p := r_p^{'} - r_p^{''} \tag{4.25}$$

The same form is valid for the vector λ:

$$\lambda(j) = \delta^{r_\lambda} \frac{b_{j,r_\lambda^{'}+1}^{\lambda} + \sum\limits_{s=1}^{m+N-r_\lambda^{'}-1} b_{j,s+r_\lambda^{'}+1}^{\lambda}\delta^{s}}{a_{j,r_\lambda^{''}+1}^{\lambda} + \sum\limits_{s=1}^{m+N} a_{j,s+r_\lambda^{''}+1}^{\lambda}\delta^{s}}, \quad r_\lambda := r_\lambda^{'} - r_\lambda^{''} \tag{4.26}$$

In view of theorem 2 (boundness of the coordinates of the saddle-point) we conclude

$$r_p \geq 0, \quad r_\lambda \geq 0$$

The assertion of this theorem follows directly from (4.25)-(4.26) and the boundness of the parameter δ. Theorem is proved. ■

The next section deals with the use of learning automata for solving the equivalent optimization problem stated above.

4.4 Optimization algorithm

It has been shown in section 4 that the stochastic constrained optimization problem (4.3)-(4.4) is asymptotically equivalent to the problem related to the determination of the saddle-points of the regularized Lagrange function $L_\delta(p, \lambda)$, using the realizations of the cost function and the constraints. This equivalent problem may be formulated and solved as the behaviour

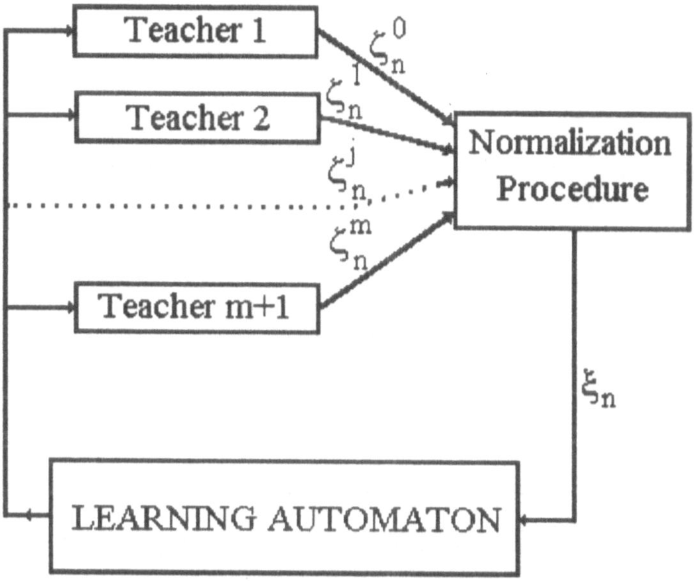

FIGURE 4.1. Multi-teacher environment.

of a variable-stochastic automaton in a multi-teacher environment [4]-[19] (Figure 4.1). Referring to the schematic block diagram for the learning automaton operating in a multi-teacher environment (Figure 4.1), we note that the normalization procedure processes as a mapping from the teachers responses ($\zeta_n^j, j = 0, ..., m$) to the learning automaton input (ξ_n).

4.4.1 Learning automata

The role of the environment (medium) is to establish the relation between the actions of the automaton and the signals received at its input. For easy reference, the mathematical description of a stochastic automaton will be given below. An automaton is an adaptive discrete machine described by:

$$\{\Xi, U, \mathcal{R}, \{\xi_n\}, \{u_n\}, \{p_n\}, T\}$$

where:

(i) Ξ is the bounded set of automaton inputs.

(ii) U denotes the set $\{u(1), u(2), ..., u(N)\}$ of actions of the automaton.

(iii) $\mathcal{R} = (\ \Omega, \mathcal{F}, \mathbf{P}\ )$ a probability space.

(iv) $\{\xi_n\}$ is a sequence of automaton inputs (environment response, $\xi_n \in \Xi$) provided by the environment in a binary (P-model environment) or continuous (S-model environment) form.

(v) $\{u_n\}$ is a sequence of automaton outputs (actions).

(vi) $p_n = [p_n(1), p_n(2), ..., p_n(N)]^T$ is the conditional probability distribution at time n

$$p_n(i) = \mathbf{P}\{\omega : u_n = u(i) \quad / \quad \mathcal{F}_{n-1}\}, \quad \sum_{i=1}^{N} p_n(i) = 1 \quad \forall n$$

where $\mathcal{F}_n = \sigma(\xi_1, u_1, p_1; ...; \xi_n, u_n, p_n)$ is the σ-algebra generated by the corresponding events ($\mathcal{F}_n \in \mathcal{F}$).

(vii) T represents the reinforcement scheme (updating scheme) which changes the probability vector p_n to p_{n+1}:

$$p_{n+1} = p_n + \gamma_n T_n(p_n; \{\xi_t \quad \}_{t=1,...,n}; \{u_t\}_{t=1,...,n}) \tag{4.27}$$

$$p_1(i) > 0 \quad \forall i = 1, ..., N$$

where γ_n is a scalar correction factor and the vector

$$T_n(.) = \left[T_n^1(.), ..., T_n^N(.)\right]^T$$

satisfies the following conditions (for preserving probability measure):

$$\sum_{i=1}^{N} T_n^i(.) = 0 \quad \forall n \tag{4.28}$$

$$p_n(i) + \gamma_n T_n^i(.) \in [0, 1] \quad \forall n, \forall i = 1, ..., N \tag{4.29}$$

This is the heart of the learning automaton.

4.4.2 Algorithm

Let $u_n = u(i)$ be the action selected at time n. For fixed vector λ_n and fixed regularization parameter $\delta = \delta_n$, the environment response (automaton input) ξ_n is defined as follows:

$$\xi_n := \frac{\alpha_n y_n + \beta_n}{p_n(i)}, \qquad u_n = u(i) \tag{4.30}$$

where

$$y_n := \zeta_n^0 + \sum_{j=1}^{m} \lambda_n(j) \zeta_n^j + \delta_n p_n(i) \tag{4.31}$$

and

$$\zeta_n^T(\omega) := \left(\zeta_n^0(u_n,\omega), \zeta_n^1(u_n,\omega), ..., \zeta_n^m(u_n,\omega)\right), \quad \zeta_n(\omega) \in R^{m+1} \tag{4.32}$$

The sequences $\{\alpha_n\}$ and$\{\beta_n\}$ are positive deterministic. They will be defined below.

The transformation (4.30)-(4.31) is a normalization procedure [4].

In this study we shall be concerned with Bush-Mosteller [15] reinforcement scheme for adjusting the probability distribution p (other reinforcement schemes can be also used [4]). The Bush-Mosteller scheme [4]-[15], is described by:

$$p_{n+1} = p_n + \gamma_n \left[e(u_n) - p_n + \xi_n(e^N - Ne(u_n))/(N-1)\right] \tag{4.33}$$

$$p_1(i) = \frac{1}{N}, \quad (i = 1, ..., N)$$

where

$$\begin{aligned} \gamma_n &\in [0,1], \quad \xi_n \in [0,1] \\ e(u_n) &= (\underbrace{0,...,0,1}_{i},0...0)^T, \quad u_n = u(i) \\ e^N &= (1,...,1)^T \in R^N \end{aligned}$$

The Lagrange multipliers λ_n are adjusted according to the following algorithm:

$$\lambda_{n+1}(j) = \left[\lambda_n(j) + \gamma_n^\lambda \left(\zeta_n^j - \delta_n \lambda_n(j)\right)\right]_0^{\lambda_n^+} \tag{4.34}$$

$$\gamma_n^\lambda \geq 0, \ \lambda_1(j) > 0 \ , j = 1, ..., m$$

where the operator $[\cdot]_0^{\lambda_n^+}$ is defined as

$$[x]_0^{\lambda_n^+} := \begin{cases} \lambda_n^+ & x \geq \lambda_n^+ \\ x & 0 \leq x \leq \lambda_n^+ \\ 0 & x \leq 0 \end{cases}$$

and $\{\lambda_n^+\}$ is a monotically increasing sequence.

The sequences α_n and β_n will be selected according to the following lemma.

Lemma 2. *If, for some positive monotically decreasing* $\{\tau_n\}$ $(0 < \tau_n \downarrow 0)$ *and positive increasing* $\{\lambda_n^+\}$ $(0 < \lambda_n^+ \uparrow \infty)$ *sequences, the parameters* α_n *and* β_n *are given by*

$$\alpha_n := \frac{\widetilde{\alpha}_n}{\sigma_0^+ + \lambda_n^+ \sum_{j=1}^{m} \sigma_j^+ + \delta_n}, \quad \beta_n := \widetilde{\alpha}_n + \frac{\xi_n^- \tau_n}{1 + (N-2)\tau_n}$$

where

$$\widetilde{\alpha}_n := \frac{\xi_n^+ \tau_n}{2\left(1+(N-2)\tau_n\right)},\ \xi_n^+ := 1-\tau_n \text{ and } \xi_n^- := \tau_n(N-1)$$

then, the automaton input belongs to $(0,1]$*:*

$$\xi_n \in \left[\frac{(N-1)\tau_n^2}{1+(N-2)\tau_n^2}; 1\right] \subset (0,1]$$

Proof.

To prove this lemma we will use the induction method. Let us assume that $\xi_{n-1} \in \left[\xi_{n-1}^-, \xi_{n-1}^+\right]$, from the Bush-Mosteller scheme (4.35), we derive

$$p_n(i) = p_{n-1}(i) + \gamma_{n-1}\left[\chi(u_{n-1} = u(i)) - p_{n-1}(i) + \right.$$

$$\left. +\xi_{n-1}[1 - N\chi(u_{n-1} = u(i))]/(N-1)\right] \geq$$

$$\geq p_{n-1}(i)\left(1-\gamma_{n-1}\right) + \gamma_{n-1}\min\left\{1-\xi_{n-1}; \frac{\xi_{n-1}}{N-1}\right\} \geq$$

$$\geq p_{n-1}(i)\left(1-\gamma_{n-1}\right) + \gamma_{n-1}\min\left\{1-\xi_{n-1}^+; \frac{\xi_{n-1}^-}{N-1}\right\} \tag{4.35}$$

From the definition of ξ_{n-1}^- and ξ_{n-1}^+, we derive

$$\tau_{n-1} = \min\left\{1-\xi_{n-1}^+; \frac{\xi_{n-1}^-}{N-1}\right\}$$

Consequently, it follows

$$p_n(i) \geq \min\ \{p_{n-1}(i); \tau_{n-1}\} \geq \min\{\min\ \{p_{n-2}(i); \tau_{n-2}\}; \tau_{n-1}\} =$$

$$= \min\ \{p_{n-2}(i); \tau_{n-1}\} \geq ... \geq \min\ \{p_1(i); \tau_{n-1}\} =$$

$$= \min\left\{\frac{1}{N}; \tau_{n-1}\right\} = \tau_{n-1}$$

Finally,

$$p_n(i) \geq \tau_{n-1}, \forall i = 1, ..., N \tag{4.36}$$

Consider now the automaton input ξ_n:

$$\xi_n = \frac{1}{(1+(N-2)\tau_n)}\left[\frac{\xi_n^+ \tau_n}{2p_n(i)}\left(\frac{y_n}{\sigma_0^+ + \lambda_n^+ \sum\limits_{j=1}^{m} \sigma_j^+ + \delta_n}\right) + \right.$$

$$+\frac{\xi_n^+\tau_n/2+\xi_n^-\tau_n}{p_n}\Bigg] \geq$$

$$\geq \frac{1}{(1+(N-2)\tau_n)}\left[\frac{\xi_n^+\tau_n/2}{p_n}(-1)+\frac{\xi_n^+\tau_n/2+\xi_n^-\tau_n}{p_n}\right]=$$

$$=\frac{\xi_n^-\tau_n}{(1+(N-2)\tau_n)\,p_n}\geq \xi_n^-\frac{\tau_n}{1+(N-2)\tau_n}=\frac{\tau_n^2(N-1)}{1+(N-2)\tau_n}=\xi_n^-$$

and

$$\xi_n \leq \frac{(\xi_n^+ + \xi_n^-)\,\tau_n}{(1+(N-2)\tau_n)\,p_n} \leq \frac{(\xi_n^+ + \xi_n^-)\,\tau_n}{(1+(N-2)\tau_n)\,\tau_{n-1}} \leq$$

$$\leq \frac{(\xi_n^+ + \xi_n^-)}{1+(N-2)\tau_n}=\frac{1+\tau_n(N-2)}{1+(N-2)\tau_n}=1$$

Lemma is proved. ■

As might be expected from the construction of the automaton input (normalized environment response), the signal input ξ_n belongs to the interval $[\xi_{n-1}^-, \xi_{n-1}^+]$.

This lemma leads to the following corollary

Corollary 1. *The conditional mathematical expectation of ξ_n is given by the affine transformation of the corresponding regularized Lagrange function (4.20):*

$$\tau_n^p(i) := \mathbf{E}\{\xi_n \mid (u_n = u(i)) \wedge \mathcal{F}_n\} = \frac{1}{p_n(i)}\left[\alpha_n \frac{\partial}{\partial p_i} L_{\delta_n}(p_n, \lambda_n) + \beta_n\right]$$

where α_n and β_n are defined in the previous lemma.

The asymptotic properties of this optimization algorithm are stated in the next section.

4.5 Convergence and convergence rate analysis

In this section we prove the convergence of the optimization algorithm described in the previous section and we estimate its convergence rate. The next theorems are our main results in this section.

Theorem 4. *Consider the optimization algorithm (4.33)-(4.34), under assumptions (**H1**) and (**H2**), and in addition, assume that there exist two nonnegative sequences $\{\varepsilon_n\}$ and $\{\tau_n\}$ such that*

1)

$$0<\varepsilon_n \downarrow 0,\ 0<\tau_n \leq \frac{1}{N+1}\downarrow 0,\ \sum_{n=1}^{\infty}\gamma_n\delta_n\tau_n\left(\lambda_n^+\right)^{-1}=\infty,$$

2) the following limit

$$d := \lim_{n\to\infty} \frac{\varepsilon_n \lambda_n^+}{\gamma_n \delta_n \tau_n} h < 1, \; h := 2\left(1 - \frac{1}{N}\right)\sum_{j=1}^{m} \sigma_j^+$$

exists,

3) the sequence $\{\gamma_n^\lambda\}$ *is constructed as follows*

$$\gamma_n^\lambda = \frac{N}{N-1}\alpha_n \gamma_n.$$

Then,

if

4)

$$\sum_{n=1}^{\infty}\left[\gamma_n^2 + \frac{|\delta_{n+1} - \delta_n|^2}{\varepsilon_n}\right] < \infty$$

then, the convergence to the saddle-point (p^{**}, λ^{**}) *(of the Lagrange function) which has the minimal norm, is ensured with probability one, i.e.,*

$$W_n := \|p_n - p_n^*\|^2 + \|\lambda_n - \lambda_n^*\|^2 \overset{a.s.}{\to} 0$$

and

$$\|p_n - p^{**}\|^2 + \|\lambda_n - \lambda^{**}\|^2 \overset{a.s.}{\to} 0,$$

if

5)

$$\frac{\gamma_n \lambda_n^+}{\delta_n \tau_n} + \frac{|\delta_{n+1} - \delta_n|^2 \lambda_n^+}{\varepsilon_n \gamma_n \delta_n \tau_n} \to 0$$

then, we obtain the mean squares convergence

$$\mathbf{E}\{W_n\} \underset{n\to\infty}{\to} 0, \left\{\|p_n - p^{**}\|^2 + \|\lambda_n - \lambda^{**}\|^2\right\} \underset{n\to\infty}{\to} 0$$

Proof.

For any time $n \geq \inf_t \{t : \|\lambda_t^*\| \leq \lambda_t^+\}$, (4.33) and (4.34) lead to

$$W_{n+1} \leq \left\| p_n + \gamma_n \left[e(u_n) - p_n + \xi_n \frac{e^N - Ne(u_n)}{N-1} \right] - p_{n+1}^* \right\|^2 +$$

$$+ \left\| \lambda_n + \gamma_n^\lambda \left(\zeta_n - \delta_n \lambda_n \right) - \lambda_{n+1}^* \right\|^2 =$$

$$= \left\| p_n - p_n^* + \gamma_n \left[e(u_n) - p_n + \xi_n \frac{e^N - Ne(u_n)}{N-1} \right] - (p_{n+1}^* - p_n^*) \right\|^2 +$$

$$+\left\|\lambda_n-\lambda_n^*+\gamma_n^\lambda\left(\overline{\zeta}_n-\delta_n\lambda_n\right)-\left(\lambda_{n+1}^*-\lambda_n^*\right)\right\|^2$$

where

$$\overline{\zeta}_n=\left(\zeta_n^1,...,\zeta_n^m\right)^T$$

According to the following well-known inequality

$$\|a+b\|^2\leq(1+\varepsilon)\|a\|^2+\left(1+\varepsilon^{-1}\right)\|b\|^2\ ,\varepsilon>0$$

which is valid for any $\varepsilon=\varepsilon_n>0$, it follows

$$W_{n+1}\leq(1+\varepsilon_n)\left\|\quad p_n-p_n^*+\gamma_n\left[e(u_n)-p_n+\xi_n\frac{e^N-Ne(u_n)}{N-1}\right]\right\|^2+$$

$$+(1+\varepsilon_n^{-1})\left\|\quad p_{n+1}^*-p_n^*\right\|^2+(1+\varepsilon_n)\left\|\lambda_n-\lambda_n^*+\gamma_n^\lambda\left(\overline{\zeta}_n-\delta_n\lambda_n\right)\right\|^2+$$
$$+(1+\varepsilon_n^{-1})\left\|\lambda_{n+1}^*-\lambda_n^*\right\|^2$$

In view of theorem 2, we derive

$$W_{n+1}\leq(1+\varepsilon_n)W_n+2\gamma_n(1+\varepsilon_n)\left((p_n-p_n^*)^T A_n^p+\frac{\gamma_n^\lambda}{\gamma_n}(\lambda_n-\lambda_n^*)^T A_n^\lambda\right)+$$
$$+\gamma_n^2(1+\varepsilon_n)\left(\|A_n^p\|^2+\left(\frac{\gamma_n^\lambda}{\gamma_n}\right)^2\left\|A_n^\lambda\right\|^2\right)+(1+\varepsilon_n^{-1})C^2\left|\delta_{n+1}-\delta_n\right|^2 \tag{4.37}$$

where

$$A_n^p:=e(u_n)-p_n+\xi_n\frac{e^N-Ne(u_n)}{N-1}$$

$$A_n^\lambda:=\overline{\zeta}_n-\delta_n\lambda_n$$

Let us calculate the conditional mathematical expectations and the estimation of the conditional second moments of the vectors A_n^p and A_n^λ. Notice that p_n and λ_n are $\mathcal{F}_n$-measurable. Hence, we have:

1. Conditional mathematical expectations

   1) Vectors A_n^p

$$\mathbf{E}\left\{A_n^p|\mathcal{F}_n\right\}=\sum_{i=1}^N\mathbf{E}\left\{A_n^p\mid(u_n=u(i))\wedge\mathcal{F}_n\right\}p_n(i)= \tag{4.38}$$

$$=\sum_{i=1}^N\left[e(u(i))-p_n+\tau_n^p(i)\xi_n\frac{e^N-Ne(u_n)}{N-1}\right]p_n(i)=$$

$$=\frac{1}{N-1}\sum_{i=1}^N\tau_n^p(i)p_n(i)\left[e^N-Ne(u(i))\right]$$

where

$$\tau_n^p(i) := \mathbf{E}\{\xi_n \mid (u_n = u(i)) \wedge \mathcal{F}_n\}$$

2) Vectors A_n^λ

$$\mathbf{E}\{A_n^\lambda | \mathcal{F}_n\} = \sum_{i=1}^{N} \mathbf{E}\{A_n^\lambda \mid (u_n = u(i)) \wedge \mathcal{F}_n\} p_n(i) = \tag{4.39}$$

$$= \sum_{i=1}^{N} \tau_n^\lambda(i) p_n(i) - \delta_n \lambda_n$$

where

$$\tau_n^\lambda(i) := \mathbf{E}\{\overline{\zeta}_n \mid (u_n = u(i)) \wedge \mathcal{F}_n\}$$

2. Estimation of the conditional second moments

1) Vectors A_n^p

$$\mathbf{E}\left\{\|A_n^p\|^2 | \mathcal{F}_n\right\} \le c_p + 2\|p_n\|^2 \tag{4.40}$$

where

$$c_p = 1 + \frac{2N}{N-1}$$

2) Vectors A_n^λ

$$\mathbf{E}\left\{\|A_n^\lambda\|^2 | \mathcal{F}_n\right\} \le 2\sigma_\zeta^2 + 2\delta_n^2 (\lambda_n^+)^2 \tag{4.41}$$

$$\sigma_\zeta^2 := \sum_{j=1}^{m} (\sigma_j^+)^2 \tag{4.42}$$

Considering the conditional mathematical expectation of both sides of (4.37) and combining (4.38),(4.39),(4.40) and (4.41) we obtain

$$\mathbf{E}\{W_{n+1}|\mathcal{F}_n\} \overset{a.s.}{\le} (1+\varepsilon_n)W_n +$$

$$+2\gamma_n(1+\varepsilon_n)\left\{(p_n - p_n^*)^T \left(\frac{1}{N-1}\sum_{i=1}^{N} \tau_n^p(i) p_n(i)\left[e^N - Ne(u(i))\right]\right) +\right.$$

$$\left. +\frac{\gamma_n^\lambda}{\gamma_n}(\lambda_n - \lambda_n^*)^T \left(\sum_{i=1}^{N} \tau_n^\lambda(i) p_n(i) - \delta_n \lambda_n\right)\right\} +$$

$$+\gamma_n^2(1+\varepsilon_n)\left(c_p + 2\|p_n\|^2 + \left(\frac{\gamma_n^\lambda}{\gamma_n}\right)^2 \left[2\sigma_\zeta^2 + 2\delta_n^2 (\lambda_n^+)^2\right]\right) +$$

$$+(1+\varepsilon_n^{-1})C^2\left|\delta_{n+1}-\delta_n\right|^2$$

Observe that

$$\mathbf{E}\left\{A_n^p|\mathcal{F}_n\right\}=\frac{1}{N-1}\sum_{i=1}^{N}\tau_n^p(i)p_n(i)\left[e^N-Ne(u(i))\right]\overset{a.s.}{=}$$

$$\overset{a.s.}{=}\frac{1}{N-1}\sum_{i=1}^{N}\left(\alpha_n\frac{\partial}{\partial p(i)}L_{\delta_n}(p_n,\lambda_n)+\beta_n\right)\left[e^N-Ne(u(i))\right] \tag{4.43}$$

$$\overset{a.s.}{=}-a_n\nabla_p L_{\delta_n}(p_n,\lambda_n)-b_n e^N$$

where

$$a_n:=\frac{N\alpha_n}{N-1},\quad b_n:=\frac{N\beta_n}{N-1} \tag{4.44}$$

and α_n and β_n are deterministic sequences, defined by lemma 2,

$$\mathbf{E}\left\{A_n^\lambda|\mathcal{F}_n\right\}=\sum_{i=1}^{N}\tau_n^\lambda(i)p_n(i)-\delta_n\lambda_n\overset{a.s.}{=}\nabla_\lambda L_{\delta_n}(p_n,\lambda_n) \tag{4.45}$$

Using formulas (4.43) and (4.45) we derive

$$\mathbf{E}\left\{W_{n+1}|\mathcal{F}_n\right\}\overset{a.s.}{\leq}(1+\varepsilon_n)(1+2\gamma_n^2)W_n+ \tag{4.46}$$

$$-2\gamma_n(1+\varepsilon_n)\left\{a_n\left(p_n-p_n^*\right)^T\nabla_p L_{\delta_n}(p_n,\lambda_n)-\right. \tag{4.47}$$

$$\left.-\frac{\gamma_n^\lambda}{\gamma_n}\left(\lambda_n-\lambda_n^*\right)^T\nabla_\lambda L_{\delta_n}(p_n,\lambda_n)\right\}+$$

$$+\gamma_n^2(1+\varepsilon_n)\left(c_p+2\left(\frac{\gamma_n^\lambda}{\gamma_n}\right)^2\left[\sigma_\varsigma^2+\delta_n^2\left(\lambda_n^+\right)^2\right]\right)+(1+\varepsilon_n^{-1})C^2\left|\delta_{n+1}-\delta_n\right|^2$$

Let

$$\gamma_n^\lambda=a_n\gamma_n$$

Then, taking into account the strictly convexity property of the regularized Lagrange function $L_{\delta_n}(p_n,\lambda_n)$ (see Lemma 1) and (4.47), we obtain

$$\mathbf{E}\left\{W_{n+1}|\mathcal{F}_n\right\}\overset{a.s.}{\leq}(1+\varepsilon_n)\left(1-\gamma_n\delta_n a_n\right)W_n+$$

$$+\gamma_n^2(1+\varepsilon_n)\left(c_p+2a_n^2\left[\sigma_\varsigma^2+\delta_n^2\left(\lambda_n^+\right)^2\right]\right)+(1+\varepsilon_n^{-1})C^2\left|\delta_{n+1}-\delta_n\right|^2$$

Finally,

$$\mathbf{E}\{W_{n+1}|\mathcal{F}_n\} \overset{a.s.}{\leq} (1+\varepsilon_n)(1-\gamma_n\delta_n a_n) W_n + \tag{4.48}$$

$$+\gamma_n^2 c_1 + \gamma_n^2 a_n^2 \delta_n^2 \left(\lambda_n^+\right)^2 c_3 + \frac{|\delta_{n+1}-\delta_n|^2}{\varepsilon_n} c_2$$

where

$$c_1 := 2\left(c_p + 2\sigma_\zeta^2 \sup_n a_n^2\right), \; c_2 := 2C^2, \; c_3 := 4 \sup_n a_n^2$$

According to assumption 2 of this theorem, it follows:

$$a_n = \frac{N + o(1)}{2(N-1)\sum_{j=1}^{m} \sigma_j^+} \left(\frac{\tau_n}{\lambda_n^+}\right), \quad \frac{\varepsilon_n}{\gamma_n \delta_n a_n} = d + o(1)$$

and hence, we conclude

$$(1+\varepsilon_n)(1-\gamma_n\delta_n a_n) \leq$$

$$\leq 1 - \gamma_n \delta_n a_n \left[1 - \frac{\varepsilon_n}{\gamma_n \delta_n a_n}\right] \leq$$

$$\leq 1 - \gamma_n \delta_n a_n \left[1 - d + o(1)\right]$$

From (4.48) we obtain the following quasimartingale inequality

$$\mathbf{E}\{W_{n+1}|\mathcal{F}_n\} \overset{a.s.}{\leq} (1 - \gamma_n\delta_n a_n \left[1 - d + o(1)\right]) W_n + \tag{4.49}$$

$$+\gamma_n^2 c_1 + c_2 \frac{|\delta_{n+1}-\delta_n|^2}{\varepsilon_n} + \gamma_n^2 a_n^2 \delta_n^2 \left(\lambda_n^+\right)^2 c_3$$

It follows that under the assumptions of this theorem, all the conditions of the Robbins-Siegmund theorem [4]-[32] (see Appendix B), are satisfied. We conclude

$$W_n \overset{a.s.}{\to} W^*, \; \sum_{n=1}^{\infty} \gamma_n \delta_n a_n W_n \overset{a.s.}{<} \infty$$

and, in view of assumption 1, $\sum_{n=1}^{\infty} \gamma_n \delta_n \tau_n (\lambda_n^+)^{-1} = \infty$, this result implies that there exists a subsequence W_{n_k} which tends to zero and, hence $W^* \overset{a.s.}{=} 0$.

The mean squares convergence follows directly from (4.49) after applying the operator of mathematical expectation and lemma A5 in [4] for $\delta_n = 0$ and $b = 0$.

Theorem is proved. ■

Corollary 2. *If*

$$\lambda_n^+ = \lambda_1 + n^\lambda \ln n,\ \tau_n = \frac{1}{(N+1)\left(1+n^\tau \ln n\right)},\ \lambda_1 > 0 \qquad (4.50)$$

$$\gamma_n = \frac{\gamma_0}{n^\gamma},\ \delta_n = \frac{\delta_0}{n^\delta},\ \varepsilon_n = \frac{\varepsilon_0}{n^\varepsilon},\ \ \varepsilon_0 < \gamma_0\lambda_1 h \qquad (4.51)$$
$$\gamma,\ \delta,\ \varepsilon,\ \tau > 0,\ \lambda \geq 0$$

then,

1) if

$$\lambda + \gamma + \tau + \delta = \varepsilon \leq 1,\ 2\gamma > 1,\ 2(\delta + 1) - \varepsilon > 1 \qquad (4.52)$$

then, conditions (1)-(4) of the previous theorem are fulfilled and, the almost surely convergence is ensured, i.e.,

$$W_n \overset{a.s.}{\to} 0$$

2) if

$$\lambda + \gamma + \tau + \delta = \varepsilon \leq 1,\ \gamma > \lambda + \tau + \delta,\ 2 + \delta > \varepsilon + \gamma + \lambda + \tau \quad (4.53)$$

then, conditions (1),(2), (3) and (5) of the previous theorem are fulfilled and, the mean squares convergence holds, i.e.,

$$\mathbf{E}\left\{W_n\right\} \to 0$$

Proof.

The proof follows directly by substituting (4.50)-(4.53) into the conditions of the previous theorem. ■

The estimation of the maximum convergence rate for the class of parameters (4.50)-(4.53) is given by the next theorem.

Theorem 5. *Under the conditions of the previous theorem and for the class of parameters (4.50)-(4.53), there exists $\nu > 0$ such that*

$$W_n \overset{a.s.}{=} o_\omega\left(\frac{1}{n^\nu}\right) \text{ and } \mathbf{E}\left\{W_n\right\} = o\left(\frac{1}{n^\nu}\right)$$

where the order ν of convergence rate satisfies the following constraint

$$\nu < \nu^*(\gamma, \varepsilon, \delta) \leq \nu^{**} = \frac{1}{3}$$

and the maximum convergence rate ν^{**} is reached for

$$\gamma = \gamma^* = \frac{2}{3},\ \varepsilon = \varepsilon^* = 1,\ \tau = \tau^* = \delta = \delta^* = \frac{1}{6},\ \lambda = \lambda^* = 0$$

$$\nu^*(\gamma^*, \varepsilon^*, \delta^*) = \nu^{**}$$

Proof.

From theorem 3, it follows:

$$W_n^* := \|p_n - p^{**}\|^2 + \|\lambda_n - \lambda^{**}\|^2 = \|(p_n - p_n^*) + (p_n^* - p^{**})\|^2 +$$
$$+ \|(\lambda_n - \lambda_n^*) + (\lambda_n^* - \lambda^{**})\|^2 \leq 2\|p_n - p_n^*\|^2 + 2\|p_n^* - p^{**}\|^2 +$$
$$+2\|\lambda_n - \lambda_n^*\|^2 + 2\|\lambda_n^* - \lambda^{**}\|^2 \leq 2W_n + C\delta_n^2$$

Multiplying both sides of the previous inequality by ν_n, we derive

$$\nu_n W_n^* \leq 2\nu_n W_n + \nu_n C\delta_n^2$$

Selecting $\nu_n = n^\nu$ and in view of lemma A.3-2 (see Appendix A) and taking into account that

$$\frac{\nu_{n+1} - \nu_n}{\nu_n} = \frac{\nu + o(1)}{n}$$

we obtain

$$0 < \nu < \min\{2\gamma - 1; 2\delta + 1 - \varepsilon, 2\delta\} := \nu^*(\gamma, \varepsilon, \delta)$$

where the positive parameters γ, δ, ε, τ,and λ satisfy the restrictions (4.5)

$$\lambda + \gamma + \tau + \delta = \varepsilon \leq 1,\ 2\gamma > 1,\ 2(\delta + 1) - \varepsilon > 1$$

and the constraint (4.36) which is equivalent to

$$\delta \leq \tau$$

Substituting

$$\varepsilon = \lambda + \gamma + \tau + \delta$$

into the previous restrictions (constraints), we finally obtain

$$\frac{1}{2} < \gamma \leq 1 - \delta - \tau - \lambda,\ \delta \leq \tau$$

and similarly it follows

$$0 < \nu < \min\{2\gamma - 1; 1 + \delta - \tau - \lambda - \gamma, 2\delta\} := \nu^*(\gamma, \varepsilon, \delta)$$

The solution of this optimization problem

$$\nu^*(\gamma, \varepsilon, \delta) \to \max_{\gamma, \varepsilon, \delta}$$

is given by

$$2\gamma - 1 = 1 + \delta - \tau - \lambda - \gamma = 2\delta$$

or in equivalent form

$$\gamma = \frac{2}{3} - \frac{1}{3}(\lambda + \tau - \delta) = \frac{1}{2} + \delta = 1 - \delta - \tau - \lambda$$

The smallest τ maximizing γ is equal to

$$\tau = \delta$$

and, hence

$$\gamma = \frac{2}{3} - \frac{1}{3}\lambda = \frac{1}{2} + \delta = 1 - 2\delta - \lambda$$

From these relations, we derive

$$\lambda = \frac{1}{2} - 3\delta$$

Taking into account that $\lambda \geq 0$, we obtain

$$\delta \leq \frac{1}{6}$$

and, consequently

$$\gamma = \frac{1}{2} + \delta \leq \frac{2}{3}$$

The optimal parameters are

$$\gamma = \gamma^* = \frac{2}{3}, \; \varepsilon = \varepsilon^* = 1, \; \tau = \tau^* = \delta = \delta^* = \frac{1}{6}, \; \lambda = \lambda^* = 0$$

The maximum convergence rate is achieved with these parameters and is equal to

$$\nu^*(\gamma, \varepsilon, \delta) = 2\gamma^* - 1 = \nu^{**} = \frac{1}{3}$$

Similarly, after application of the mathematical expectation operator to (4.49), for $\mathbf{E}\{W_n\}$ and in view of lemma A.3-2 (see appendix A) and lemma A5 in [4], we obtain the desired results.

Theorem is proved. ■

An increasing attention has been devoted to the use of penalty functions for the solution of nonlinear programming problems. The next part of this chapter deals with the penalty function approach.

4.6 Penalty function

Consider the following programming problem

$$V_0(p) = \sum_{i=1}^{N} v_i^0 p(i) \to \min_{p \in S^N} \tag{4.54}$$

$$V_j(p) = \sum_{i=1}^{N} v_i^j p(i) \leq 0 \ (j = 1, ..., m) \tag{4.55}$$

Let us assume that the set P_r of all vectors $p \in S^N$ satisfying the constraints (4.55) is not empty, i.e.,

$$P_r := p \in S^N \mid V_j(p) \leq 0 \ (j = 1, ..., m) \neq 0 \tag{4.56}$$

This problem (4.54-4.55) is equivalent [10] to a linear programming problem with equality constraints

$$V_0(p) \rightarrow \min_{p \in S^N, \widetilde{u} \in R_+^m} \tag{4.57}$$

$$V_j(p) + \widetilde{u}_j = 0, \ \widetilde{u}_j \geq 0 \ (j = 1, ..., m) \tag{4.58}$$

where $\widetilde{u}_j$ are the nonnegative slack variables [26]-[27].

In this part we shall be concerned with the penalty function approach. The following penalty function has been considered by Poznyak [18]

$$V_0(p) + \mu \sum_{i=1}^{m} \left([V_j(p) + \widetilde{u}_j]^+ \right)^2, \ \mu > 0 \tag{4.59}$$

where the penalty coefficient $\mu \rightarrow \infty$.

The optimal solution of this penalty function is equivalent to the optimal solution of the following penalty function

$$\mu V_0(p) + \sum_{i=1}^{m} \left([V_j(p) + \widetilde{u}_j]^+ \right)^2, \ 0 < \mu \rightarrow 0 \tag{4.60}$$

This penalty function is more useful (practical point of view) because $\mu \rightarrow 0$ instead to infinity.

Let us consider the following regularized penalty function [24]

$$L_{\mu,\delta}(p, \widetilde{u}) := \mu V_0(p) + \frac{\delta}{2} \left(\|p\|^2 + \|\widetilde{u}\|^2 \right) + \frac{1}{2} \|Vp + \widetilde{u}\|^2 \tag{4.61}$$

where

$$\begin{gathered} V = \left[\overline{V}_1 \vdots \ldots \vdots \overline{V}_N \right] \in R^{m+N} \\ \overline{V}_j^T := \left(v_1^j, ..., v_N^j \right) \in R^N \ (j = 1, ..., m) \\ \widetilde{u}^T := (\widetilde{u}_1, ..., \widetilde{u}_m) \in R^m \end{gathered} \tag{4.62}$$

We assume that the parameters μ (penalty parameter) and δ (regularization parameter) are positive and tend monotically to zero, i.e.,

$$0 < \mu, \delta \downarrow 0$$

The parameter δ play the same role as in the first part.

We shall be concerned by the following optimization problem

$$L_{\mu,\delta}(p,\widetilde{u}) \to \inf_{p\in R^N,\widetilde{u}\in R_+^m} \tag{4.63}$$

For fixed $\mu > 0$ and $\delta > 0$, this problem is strictly convex. Its solution is unique and will be denoted by

$$(p^*(\mu,\delta),\widetilde{u}^*(\mu,\delta)) \in S^N \times R_+^m \tag{4.64}$$

The properties of the optimal solution are presented in the next section.

4.7 Properties of the optimal solution

The next theorem states the property of this solution (4.64) when δ and μ tend to zero.

Theorem 6. *Let us assume that*

1) the set P_r is not empty and the Slater's condition

$$V_j(\overline{p}) > 0 \ (j = 0, ..., m) \tag{4.65}$$

is fulfilled for some $\overline{p} \in S^N$

2) The parameters μ and δ are time-varying, i.e.,

$$\mu = \mu_n, \ \delta = \delta_n, \ (n = 1, 2, ...)$$

such that

$$0 < \mu \downarrow 0, \ 0 < \delta \downarrow 0 \tag{4.66}$$

$$\frac{\delta_n}{\mu_n} \underset{n\to\infty}{\to} 0, \ \frac{(\mu_n)^{\frac{3}{2}}}{\delta_n} \underset{n\to\infty}{\to} 0 \tag{4.67}$$

Then,

$$\begin{aligned} p_n^* &:= p^*(\mu_n, \delta\ ) \underset{n\to\infty}{\to} p^{**} \\ \widetilde{u} &:= \widetilde{u}(\mu_n, \delta_n) \underset{n\to\infty}{\to} \widetilde{u}^{**} \end{aligned} \tag{4.68}$$

where $p^{**} \in P_r$ is the solution with the minimal weighted norm, of the linear programming problem (4.54), i.e.,

$$\|p^{**}\|_{I+V^TV} \le \|p^*\|_{I+V^TV}, \ p^* \in P^* \subseteq P_r \tag{4.69}$$

where p^* is any solution of (4.54), P^* is the set of all solutions and, u^{**} is given by

$$\widetilde{u}^{**} = -Vp^{**} \tag{4.70}$$

and $\|x\|_Q^2 := x^TQx$.

Proof.

First, Let us prove [24] that the Hessian matrix associated with the penalty function (4.61) is strictly positive definite for any positive μ and δ, i.e.,

$$H := \begin{bmatrix} \nabla_p^2 L_{\mu,\delta}(p,\tilde{u}) & \nabla_p \nabla_{\tilde{u}}^T L_{\mu,\delta}(p,\tilde{u}) \\ \nabla_{\tilde{u}} \nabla_p^T L_{\mu,\delta}(p,\tilde{u}) & \nabla_{\tilde{u}}^2 L_{\mu,\delta}(p,\tilde{u}) \end{bmatrix} > 0 \quad (4.71)$$

To prove this result it is sufficient to show that the following matrix are strictly definite positive [33]

$$\nabla_p^2 L_{\mu,\delta}(p,\tilde{u}) > 0,\ \nabla_{\tilde{u}}^2 L_{\mu,\delta}(p,\tilde{u}) > 0,$$

$$\nabla_p^2 L_{\mu,\tilde{u}}(p,\tilde{u}) - \nabla_p \nabla_{\tilde{u}}^T L_{\mu,\delta}(p,\tilde{u}) \left[\nabla_{\tilde{u}}^2 L_{\mu,\delta}(p,\tilde{u})\right]^{-1} \nabla_{\tilde{u}} \nabla_p^T L_{\mu,\delta}(p,\tilde{u}) > 0$$

$$\nabla_{\tilde{u}}^2 L_{\mu,\tilde{u}}(p,\tilde{u}) - \nabla_{\tilde{u}} \nabla_p^T L_{\mu,\delta}(p,\tilde{u}) \left[\nabla_p^2 L_{\mu,\tilde{u}}(p,\tilde{u})\right]^{-1} \nabla_p \nabla_{\tilde{u}}^T L_{\mu,\delta}(p,\tilde{u}) > 0$$

Indeed,

1)

$$\nabla_p^2 L_{\mu,\delta}(p,\tilde{u}) = V^T V + \delta I_N > 0,\ I_N := diag\{1,..,1\} \in R^N$$

2)

$$\nabla_{\tilde{u}}^2 L_{\mu,\delta}(p,\tilde{u}) = (1+\delta) I_m > 0$$

3)

$$\begin{aligned} &\nabla_p^2 L_{\mu,\tilde{u}}(p,\tilde{u}) - \nabla_p \nabla_{\tilde{u}}^T L_{\mu,\delta}(p,\tilde{u}) \left[\nabla_{\tilde{u}}^2 L_{\mu,\delta}(p,\tilde{u})\right]^{-1} \nabla_{\tilde{u}} \nabla_p^T L_{\mu,\delta}(p,\tilde{u}) = \\ &= V^T V + \delta I_N - \tfrac{1}{(1+\delta)} V^T I_m V = \delta I_N + \tfrac{\delta}{(1+\delta)} V^T V > 0 \end{aligned}$$

4)

$$\begin{aligned} &\nabla_{\tilde{u}}^2 L_{\mu,\tilde{u}}(p,\tilde{u}) - \nabla_{\tilde{u}} \nabla_p^T L_{\mu,\delta}(p,\tilde{u}) \left[\nabla_p^2 L_{\mu,\tilde{u}}(p,\tilde{u})\right]^{-1} \nabla_p \nabla_{\tilde{u}}^T L_{\mu,\delta}(p,\tilde{u}) = \\ &= (1+\delta) I_m - V \left(V^T V + \delta I_N\right)^{-1} V^T \end{aligned}$$

In view of the matrix inversion lemma, it follows

$$\begin{aligned} &(1+\delta) I_m - V\left(V^T V + \delta I_N\right)^{-1} V^T = (1+\delta) I_m - \\ &-\delta^{-1} V \left(I_N - V^T \left(\delta I_m + V V^T\right)^{-1} V\right) V^T = \\ &= \delta \left[I_m + \left(\delta I_m + V V^T\right)^{-1}\right] > 0\ \forall \delta > 0 \end{aligned}$$

From (4.71) it follows that the penalty function (4.61) is strictly convex and, as a consequence it has a unique minimal point on $S^N \times R_+^m$.

Let us denote it by

$$\left(p^*(\mu,\delta), \tilde{u}^*(\mu,\delta)\right) \in S^N \times R_+^m$$

Using the strictly convex property, we conclude that

$$0 \leq \begin{bmatrix} p - p^*(\mu,\delta) \\ \widetilde{u} - \widetilde{u}^*(\mu,\delta) \end{bmatrix}^T H \begin{bmatrix} p - p^*(\mu,\delta) \\ \widetilde{u} - \widetilde{u}^*(\mu,\delta) \end{bmatrix} \leq$$

$$\leq (p - p^*(\mu,\delta))^T \nabla_p L_{\mu,\delta} (p^*(\mu,\delta), \widetilde{u}^*(\mu,\delta)) +$$

$$+ (\widetilde{u} - \widetilde{u}^*(\mu,\delta))^T \nabla_u L_{\mu,\delta} (p^*(\mu,\delta), \widetilde{u}^*(\mu,\delta)) = \tag{4.72}$$

$$= (p - p^*(\mu,\delta))^T \left(\mu\overline{v}_0 + \delta p^*(\mu,\delta) + V^T [V p^*(\mu,\delta) + \widetilde{u}^*(\mu,\delta)]\right) +$$

$$+ (\widetilde{u} - \widetilde{u}^*(\mu,\delta))^T (V p^*(\mu,\delta) + (1+\delta)\widetilde{u}^*(\mu,\delta))$$

Let p^* is any solution of the linear programming problem (4.54)-(4.55). Dividing by μ and replacing p and $\widetilde{u}$ respectively by p^* and $\widetilde{u} = -Vp^* - \delta V(p^* - p^*(\mu,\delta))$, we obtain

$$0 \leq \frac{1}{\mu}[V_0(p^*) - V_0(p^*(\mu,\delta))] + \frac{\delta}{\mu}(p^* - p^*(\mu,\delta))^T p^*(\mu,\delta) -$$

$$-\frac{1}{\mu}\|Vp^*(\mu,\delta) + \widetilde{u}^*(\mu,\delta)\|^2 + \frac{\delta}{\mu}(Vp^* - Vp^*(\mu,\delta))^T Vp^*(\mu,\delta) \leq \tag{4.73}$$

$$\leq -\frac{1}{\mu}\|Vp^*(\mu,\delta) + \widetilde{u}^*(\mu,\delta)\|^2 + \frac{\delta}{\mu}(p^* - p^*(\mu,\delta))^T \left[I_N + V^T V\right] p^*(\mu,\delta)$$

For $0 < \mu = \mu_n \downarrow 0$, $0 < \delta = \delta_n \downarrow 0$, and taking into account assumption (4.67), we derive

$$0 \geq \overline{\lim_{n\to\infty}} \frac{1}{\mu_n} \|Vp^*(\mu_n,\delta_n) + \widetilde{u}^*(\mu_n,\delta_n)\|^2 \tag{4.74}$$

and, as $0 < \mu_n \downarrow 0$ it follows

$$\|Vp^*(\mu_n,\delta_n) + \widetilde{u}^*(\mu_n,\delta_n)\|^2 \underset{n\to\infty}{\rightarrow} 0 \tag{4.75}$$

Let us denote by p^*_∞ any partial limit of the sequence $\{p^*(\mu_n,\delta_n)\}$. In view of (4.75), it follows that any partial limit $\widetilde{u}^*_\infty$ of the sequence $\{\widetilde{u}^*(\mu_n,\delta_n)\}$ is uniquely defined by

$$\widetilde{u}^*_\infty = -Vp^*_\infty$$

As $p^*_\infty \in S^N$, we derive

$$\|\widetilde{u}^*_\infty\| \leq \|Vp^*_\infty\| \leq \|V\| < \infty$$

i.e., $\{\widetilde{u}^*(\mu_n,\delta_n)\}$ is a bounded sequence.

Let us now prove that

$$\|p^*(\mu_n,\delta_n) - \pi_{P^*}\{p^*(\mu_n,\delta_n)\}\| \leq C\sqrt{\mu_n} \tag{4.76}$$

where $\pi_{P^*}\{\cdot\}$ is the projection operator onto the set P^*. It has the following property

$$\|x - \pi_{P^*}\{x\}\| \leq \|x - p^*\| \ \forall x \in R^N, \ \forall p^* \in P^* \subset R^N \tag{4.77}$$

Inequality (4.74) leads to

$$\|Vp^*(\mu_n, \delta_n) + \widetilde{u}^*(\mu_n, \delta_n)\|^2 \leq C_1\mu_n, \ C_1 \in (0, \infty)$$

Hence, because each component of the optimal solution $\widetilde{u}^*(\mu_n, \delta_n)$ is non-negative, we derive

$$Vp^*(\mu_n, \delta_n) \leq \sqrt{C_1}\sqrt{\mu_n}e^m - \widetilde{u}^*(\mu_n, \delta_n) \leq \sqrt{C_1}\sqrt{\mu_n}e^m$$
$$(e^m := (1, ..., 1)^T \in R^m)$$

This inequality must be understandable in componentwise sense.

Notice that

$$\|p^*(\mu_n, \delta_n) - \pi_{P^*}\{p^*(\mu_n, \delta_n)\}\|^2 := \min_{p \in S^N : Vp \leq 0} \|p^*(\mu_n, \delta_n) - p\|^2 \leq$$
$$\leq \max_{q \in S^N : Vq \leq \sqrt{C_1}\sqrt{\mu_n}e^m} \ \min_{p \in S^N : Vp \leq 0} \|q - p\|^2 := g(\mu_n) \tag{4.78}$$

Let us introduce the following sequence

$$\psi_n := \frac{\sqrt{C_1}\sqrt{\mu_n}}{\sqrt{C_1}\sqrt{\mu_n} + \max\limits_{j=1,...,m} V_j(\overline{p})}$$

where $\overline{p}$ satisfies the Slater's condition [34]-[35]

Hence,

$$\psi_n \in (0, 1) \tag{4.79}$$

Let us consider the following transformation

$$\widetilde{p} = (1 - \psi_n)q + \psi_n\overline{p} \in S^N \tag{4.80}$$

which transforms the set

$$\left\{q \in S^N : Vq \leq \sqrt{C_1}\sqrt{\mu_n}e^m\right\}$$

into the set

$$\{\widetilde{p} \in S^N : V\widetilde{p} \leq 0\}$$

Indeed,

$$V\widetilde{p} = (1 - \psi_n)Vq + \psi_n V\overline{p} \leq (1 - \psi_n)\sqrt{C_1}\sqrt{\mu_n}e^m + \psi_n V(\overline{p}) =$$

$$= \psi_n \left[\max_{j=1,...,m} V_j(\overline{p})e^m + V(\overline{p})\right] \leq 0$$

$$g(\mu_n) = \max_{\widetilde{p} \in S^N : V\widetilde{p} \leq 0} \min_{p \in S^N : Vp \leq 0} \left\| \frac{\widetilde{p} - \psi_n \overline{p}}{1 - \psi_n} - p \right\|^2 \leq$$

$$\leq \max_{\widetilde{p} \in S^N : V\widetilde{p} \leq 0} \left\| \frac{\widetilde{p} - \psi_n \overline{p}}{1 - \psi_n} - \widetilde{p} \right\|^2 = \left(\frac{\psi_n}{1 - \psi_n} \right)^2 \max_{\widetilde{p} \in S^N : V\widetilde{p} \leq 0} \|\overline{p} - \widetilde{p}\|^2 =$$

$$= Const \left(\frac{\psi_n}{1 - \psi_n} \right)^2 \leq Const \, (\psi_n)^2$$

from which follows the upper estimation (4.76)

Dividing both sides of (4.73) by $\frac{\delta}{\mu}$, we obtain

$$0 \geq (p^*(\mu_n, \delta_n) - p^*)^T \left[I + V^T V \right] p^*(\mu_n, \delta_n) +$$

$$+ \frac{1}{\delta_n} \|V p^*(\mu_n, \delta_n) + \widetilde{u}^*(\mu_n, \delta_n)\|^2 \geq$$

$$\geq (p^*(\mu_n, \delta_n) - p^*)^T \left[I + V^T V \right] p^*(\mu_n, \delta_n) \geq$$

$$\geq (p^*(\mu_n, \delta_n) - p^*)^T \left[I + V^T V \right] p^*(\mu_n, \delta_n) + \frac{\mu_n}{\delta_n} \overline{v}_0^T (p^*(\mu_n, \delta_n) - p^*) =$$

$$= (p^*(\mu_n, \delta_n) - p^*)^T \left[I + V^T V \right] p^*(\mu_n, \delta_n) +$$

$$+ \frac{\mu_n}{\delta_n} \overline{v}_0^T (p^*(\mu_n, \delta_n) - \pi_{P^*} \{p^*(\mu_n, \delta_n)\}) +$$

$$+ \frac{\mu_n}{\delta_n} \overline{v}_0^T (\pi_{P^*} \{p^*(\mu_n, \delta_n)\} - p^*) \geq$$

$$\geq (p^*(\mu_n, \delta_n) - p^*)^T \left[I + V^T V \right] p^*(\mu_n, \delta_n) +$$

$$+ \frac{\mu_n}{\delta_n} \overline{v}_0^T (p^*(\mu_n, \delta_n) - \pi_{P^*} \{p^*(\mu_n, \delta_n)\})$$

where $\overline{v}_0^T := (v_1^0, ..., v_N^0)$.

From (4.76), it follows

$$0 \geq (p^*(\mu_n, \delta_n) - p^*)^T \left[I + V^T V \right] p^*(\mu_n, \delta_n) - \frac{(\mu_n)^{\frac{3}{2}}}{\delta_n} \|\overline{v}_0\| C$$

For $n \to \infty$ and in view of assumption (4.67), we finally obtain

$$0 \geq (p_\infty^* - p^*)^T \left[I + V^T V \right] p_\infty^* \tag{4.81}$$

for any partial limit p_∞^*

This inequality can be interpreted as the necessary condition of optimality of the strictly convex function $f(x) := \frac{1}{2} x^T \left(I_N + V^T V \right) x$ on the convex compact set P^* [26]:

$$(x - x^*)^T \nabla f(x^*) \geq 0 \; \forall x \in P^*$$
$$x := p^*, \; x^* = \arg_{x \in P^*} f(x)$$

But any strictly convex function has a unique minimum. It follows that

$$x^* = \arg_{x \in P^*} f(x) = p^*_\infty$$

is unique (all partial limits of the sequence $\{p^*(\mu_n, \delta_n)\}$ coincide). Because $f(p^*)$ is equal to the weighted norm of the vector $p^* \in P^*$, we conclude that

$$p^*_\infty = p^{**}$$

Theorem is proved. ■

The next theorem states the Lipschitzian property of the optimal solution of the penalty function (4.61) with respect to the parameters μ and δ.

Theorem 7. *Under the assumptions of theorem 6, there exist two positive constant C_1 and C_2 such that*

$$\|p^*(\mu_1, \delta_1) - p^*(\mu_2, \delta_2)\| \le C_1 |\mu_1 - \mu_2| + C_2 |\delta_2 - \delta_1| \qquad (4.82)$$

Proof.

Similarly to the proof of Theorem 3, let us introduce the following set

$$Z_0 := \left\{ (p, \tilde{u}) : \sum_{i=1}^{N} p(i) = 1 \right\}$$
$$Z_1(i_1, ..., i_s) := \{(p, \tilde{u}) : p(i_k) = 0,\ k = 1, ..., s\} \cap Z_0$$
$$Z_2(j_1, ..., jr) := \{(p, \tilde{u}) : \tilde{u}(j_k) = 0,\ k = 1, ..., r\} \cap Z_0$$
$$Z_3(i_1, ..., i_s; j_1, ..., jr) := Z_1(i_1, ..., i_s) \cap Z_2(j_1, ..., jr)$$

The total number of these sets is equal to $2^m (2^N - 1)$. They will be denoted by $\mathcal{G}_k$ $(k = 1, ..., 2^m (2^N - 1))$. Let us associate with each set $\mathcal{G}_k$ the problem $\mathcal{B}_k$ of the optimization of the penalty function $L_{\mu,\delta}(p, u)$ (4.61) on the set $\mathcal{G}_k$. It is clear that the optimal solution of the initial problem (4.57-4.58) coincides with the solution of one of these problems $(\mathcal{B}_k)$, when $\mu, \delta \to 0$. Notice that each problem $(\mathcal{B}_k)$ concerns the optimization of the convex function $L_{\mu,\delta}(p, \tilde{u})$ (4.61) subject to equality constraints. Hence, we can use the Lagrange multipliers technique. The necessary optimality conditions are

$$\nabla_p \mathcal{L}_{\mu,\delta}(p, \tilde{u}, \lambda) = 0,\ \nabla_u \mathcal{L}_{\mu,\delta}(p, \tilde{u}, \lambda) = 0,\ \nabla_\lambda \mathcal{L}_{\mu,\delta}(p, \tilde{u}, \lambda) = 0$$

where

$$\mathcal{L}_{\mu,\delta}(p, \tilde{u}, \lambda) := L_{\mu,\delta}(p, \tilde{u}, \lambda) - \lambda_0 \left(\sum_{i=1}^{N} p(i) - 1 \right) - \sum_k \lambda_{1k} \tilde{u}(i_k) - \sum_k \lambda_{2k} p(j_k)$$

These equations can be rewritten as follows:

$$\mu \overline{V}_0 + \delta p + V^T [Vp + \tilde{u}] - \lambda_0 e^N - \sum_k \lambda_{2k} e(j_k) = 0$$

$$Vp + (1+\delta)\,\widetilde{u} - \sum_k \lambda_{1k} e(i_k) = 0 \tag{4.83}$$

$$(p, \widetilde{u}) \in \mathcal{G}_k$$

This algebraic system of equations can be written in the following compact form

$$\begin{bmatrix} V^T V + \delta I_N & V^T \\ V & (1+\delta) I_m \end{bmatrix} \begin{pmatrix} p \\ \widetilde{u} \end{pmatrix} = \begin{pmatrix} \mu \overline{V}_0 - \lambda_0 e^N - \sum_k \lambda_{2k} e(j_k) \\ \sum_k \lambda_{1k} e(i_k) \end{pmatrix}$$

or in the following equivalent polynomial form

$$p^*(\mu,\delta) \; or \; \widetilde{u}(\mu,\delta) = \left(\frac{\sum_{r=0}^{m+N} \delta^r \left(c_r^i + d_r^i \mu \right)}{\sum_{r=0}^{m+N} \delta^r h_r^i} \right)_{i=1,\dots,m+N} \tag{4.84}$$

Let us assume that the r_i'–first coefficients c_r^i and d_r^i of the numerator of (4.84) and, the r_i''–first coefficients h_r^i of the denominator of (4.84) are equal to zero. We can rewrite (4.84) in the following form:

$$p^*(\mu,\delta) \; or \; \widetilde{u}(\mu,\delta) =$$

$$= \left(\delta^{r_i} \frac{c^i_{r'+1} + d^i_{r'+1}\mu + \sum_{s=1}^{m+N-r_i'-1} \delta^s \left(c^i_{r_i'+s} + d^i_{r_i'+s}\mu \right)}{h^i_{r_i+1} + \sum_{s=1}^{m+N-r_i''-1} \delta^r h^i_{r_i''+s}} \right)_{i=\overline{1,m+N}}$$

where

$$r_i := r_i' - r_i''$$

In theorem 1 we have proved that $p^*(\mu,\delta)$ and $\widetilde{u}(\mu,\delta)$ are bounded for any μ and $\delta \to 0$. It follows:

$$r_i \geq 0 \;\; (i = 1, \dots, m+N)$$

from which the Lipschitzian property (4.82) follows. Theorem is proved. ■

The learning optimization algorithm and its properties (convergence and convergence rate) will be presented in the next section.

4.8 Optimization algorithm

The optimization problem stated above will be solved using a learning automaton operating in a multi-teacher environment. We shall be again concerned with the Bush-Mosteller reinforcement scheme [15].

$$p_{n+1} = p_n + \gamma_n \left[e(u_n) - p_n + \xi_n (e^N - Ne(u_n))/(N-1) \right] \tag{4.85}$$

$$p_1(i) = \frac{1}{N}, \quad (i = 1, ..., N)$$

where

$$\begin{aligned} \gamma_n &\in [0,1], \quad \xi_n \in [0,1] \\ e(u_n) &= (\underbrace{0, ..., 0, 1}_{i}, 0...0)^T, \quad u_n = u(i) \\ e^N &= (1, ..., 1)^T \in R^N \end{aligned}$$

The automaton input ξ_n (multi-teacher environment response) is constructed as follows:

$$\xi_n = \frac{\alpha_n y_n + \beta_n}{p_n(i)} \tag{4.86}$$

where

$$y_n = \mu_n \zeta_n^0 + \delta_n p_n(i) + \overline{\zeta}_n^T (V_{n-1} p_n + \widetilde{u})$$

$$\alpha_n = \frac{\widetilde{\alpha}_n}{\mu_n \sigma_0^+ + \delta_n + \sqrt{\sum_{j=1}^{m} (\sigma_j^+)^2} \left(\widetilde{u}_n^+ + \sqrt{N \sum_{j=1}^{m} (\sigma_j^+)^2} \right)}$$

$$\beta_n = \widetilde{\alpha}_n + \frac{\xi_n^- \tau_n}{1 + (N-2)\tau_n}, \quad \widetilde{\alpha}_n = \frac{\xi_n^+ \tau_n}{2(1 + (N-2)\tau_n)}$$

$$\xi_n^+ = 1 - \tau_n, \ \xi_n^- = \tau_n (N-1), \ 0 < \tau_n \downarrow 0, \ \widetilde{u}_n^+ \uparrow \infty$$

and V_{n-1} is the estimation of the matrix V (4.62) which elements $(V_{n-1})_{ij}$ are constructed according to the following recurrent scheme:

$$(V_n)_{ij} = (V_{n-1})_{ij} \left[1 - \frac{\chi(u_n = u(i))}{\sum_{t=1}^{n} \chi(u_t = u(i))} \right] + \frac{\chi(u_n = u(i))}{\sum_{t=1}^{n} \chi(u_t = u(i))} \zeta_n^j$$

$$(i = 1, ..., N; j = 1, ..., m)$$

In view of lemma 2 it follows that

$$\xi_n \in \left[\frac{(N-1)(\tau_n)^2}{1 + (N-2)(\tau_n)^2}; 1 \right] \in (0, 1]$$

The slack variables $\widetilde{u}_n$ will be adapted according to the following algorithm

$$\begin{gathered} \widetilde{u}_{n+1}(j) = \left[\widetilde{u}_n(j) - \gamma_n^u \left[\zeta_n^j + (1 + \delta_n) \widetilde{u}_n(j) \right] \right]_0^{u_n^+} \\ \gamma_n^u \geq 0, \ \widetilde{u}_1(j) > 0 \ (j = 1, ..., m) \end{gathered} \tag{4.87}$$

The convergence properties of this optimization algorithm are stated by the following theorem.

Theorem 8. *Consider the optimization algorithm (4.85), (4.86) and (4.87) subject to assumptions (H1) and (H2) and let*

1) the sequence $\{\gamma_n^u\}$ *be selected as follows:*

$$\gamma_n^u = \frac{N}{N-1}\alpha_n\gamma_n \tag{4.88}$$

2)

$$\overline{\lim_{n\to\infty}} \frac{2\widetilde{u}_n^+}{\gamma_n\tau_n\delta_n}\left|\frac{\alpha_{n-1}}{\alpha_n}-1\right| := d < 1 \tag{4.89}$$

3) there exist a positive sequence $\tau_n \downarrow 0$ *such that*

$$\sum_{n=1}^{\infty}\gamma_n\tau_n\delta_n\left(\widetilde{u}_n^+\right)^{-1} = \infty \tag{4.90}$$

then, if

4)

$$\sum_{n=1}^{\infty}\left[\gamma_n\tau_n\left(\widetilde{u}_n^+\right)^{-1}\frac{1}{\sqrt{n}} + \overline{s}_n + v_n\right] < \infty \tag{4.91}$$

where

$$v_n = \sqrt{\frac{\gamma_{n-1}}{\gamma_{n-1}^u}}\left[|\mu_n-\mu_{n+1}| + |\delta_n-\delta_{n+1}|\right]$$

and

$$\overline{s}_n = \gamma_n^2 + \frac{\gamma_{n-1}}{\gamma_{n-1}^u}\left[|\mu_n-\mu_{n+1}|^2 + |\delta_n-\delta_{n+1}|^2\right] + \frac{\gamma_{n-1}}{\gamma_{n-1}^u}\left(\gamma_n^u\right)^2\left(\widetilde{u}_n^+\right)^2 +$$

$$+\left|\frac{\gamma_n}{\gamma_n^u}-\frac{\gamma_{n-1}}{\gamma_{n-1}^u}\right|\left[\left(\gamma_n^u\right)^2\left(\widetilde{u}_n^+\right)^2 + |\mu_n-\mu_{n+1}|^2 + |\delta_n-\delta_{n+1}|^2\right]$$

then,

$$W_n := \|p_n - p_n^*\|^2 + \frac{\gamma_{n-1}}{\gamma_{n-1}^u}\|\widetilde{u}_n - \widetilde{u}_n^*\|^2 \underset{n\to\infty}{\overset{a.s.}{\to}} 0$$

where

$$p_n^* := p^*(\mu_n,\delta_n),\ \widetilde{u}_n^* := -Vp_n^*$$

then, if

5)

$$\frac{1}{\delta_n\sqrt{n}} + \frac{\overline{s}_n\widetilde{u}_n^+}{\gamma_n\tau_n\delta_n} + \frac{v_n\widetilde{u}_n^+}{\gamma_n\tau_n\delta_n} \underset{n\to\infty}{\to} 0 \tag{4.92}$$

then,

$$E\{W_n\} \underset{n\to\infty}{\to} 0$$

Proof.

The optimization algorithm and the properties of the projection operator lead to

$$W_{n+1} \leq \left\| p_n + \gamma_n \left[e(u_n) - p_n + \xi_n \frac{1 - Ne(u_n)}{N-1} \right] - p^*_{n+1} \right\|^2 +$$

$$+ \frac{\gamma_n}{\gamma^u_n} \left\| \widetilde{u}_n - \gamma^u_n \left[\zeta_n + (1+\delta_n) \widetilde{u}_n \right] - \widetilde{u}^*_{n+1} \right\|^2 =$$

$$= \|p - p^*_n\|^2 + \gamma_n^2 \|A^p_n\|^2 + \|p^*_n - p^*_{n+1}\|^2 +$$

$$+2\gamma_n (p - p^*_n) A^p_n + \frac{\gamma_{n-1}}{\gamma^u_{n-1}} \left[\|\widetilde{u}_n - \widetilde{u}^*_n\|^2 + (\gamma^u_n)^2 \|A^u_n\|^2 + \right.$$

$$+ \left\|\widetilde{u}^*_n - \widetilde{u}^*_{n+1}\right\|^2 - 2\gamma^u_n (\widetilde{u}_n - \widetilde{u}^*_n)^T A^u_n + 2 (\widetilde{u}_n - \widetilde{u}^*_n)^T (\widetilde{u}^*_n - \widetilde{u}^*_{n+1}) -$$

$$\left. -2\gamma^u_n (\widetilde{u}^*_n - \widetilde{u}^*_{n+1})^T A^u_n \right] + \left(\frac{\gamma_n}{\gamma^u_n} - \frac{\gamma_{n-1}}{\gamma^u_{n-1}} \right) \left\| \widetilde{u}_n - \gamma^u_n A^u_n - \widetilde{u}^*_{n+1} \right\|^2$$

where

$$A^p_n \quad : \quad = e(u_n) - p_n + \xi_n \frac{1 - Ne(u_n)}{N-1}$$

$$A^u_n \quad : \quad = \zeta_n + (1+\delta_n) \widetilde{u}_n$$

From theorem 7 it follows that

$$W_{n+1} \leq W_n + Cv_n \sqrt{W_n} + 2\gamma_n (p - p^*_n) A^p_n - 2\frac{\gamma_{n-1}}{\gamma^u_{n-1}} \gamma^u_n (\widetilde{u}_n - \widetilde{u}^*_n)^T A^u_n + \bar{s}_n \tag{4.93}$$

where

$$\bar{s}_n = 4\gamma_n^2 + 2\left(1 + \frac{\gamma_{n-1}}{\gamma^u_{n-1}}\right) \left[C_1^2 |\mu_n - \mu_{n+1}|^2 + C_2^2 |\delta_n - \delta_{n+1}|^2 \right] +$$

$$+2\frac{\gamma_{n-1}}{\gamma^u_{n-1}} \left(C_1 |\mu_n - \mu_{n+1}| + C_2 |\delta_n - \delta_{n+1}| \right) \left(C_3 + \gamma^u_n \sqrt{\theta_n} \right) + \tag{4.94}$$

$$+ \frac{\gamma_{n-1}}{\gamma^u_{n-1}} (\gamma^u_n)^2 \theta_n + 2 \left| \frac{\gamma_n}{\gamma^u_n} - \frac{\gamma_{n-1}}{\gamma^u_{n-1}} \right| \left((\gamma^u_n)^2 \theta_n + \|\widetilde{u}_n - \widetilde{u}^*_{n+1}\|^2 \right)$$

and

$$\theta_n := \sum_{j=1}^{m} (\sigma^+_j)^2 + m (\widetilde{u}^+_n)^2, \; C_3 > 0$$

Notice that

$$\|\widetilde{u}_n - \widetilde{u}^*_n\|^2 \leq \frac{\gamma^u_n}{\gamma_n} W_n$$

$$\left\|\widetilde{u}_n - \widetilde{u}^*_{n+1}\right\|^2 \leq 2 \|\widetilde{u}_n - \widetilde{u}^*_n\|^2 + 2 \left\|\widetilde{u}^*_n - \widetilde{u}^*_{n+1}\right\|^2 \leq$$

$$\leq 2\frac{\gamma_n^u}{\gamma_n}W_n + 2\left[C_1^2\left|\mu_n - \mu_{n+1}\right|^2 + C_2^2\left|\delta_n - \delta_{n+1}\right|^2\right]$$

Using (4.41)-(4.43) we derive:

$$\mathbf{E}\left\{A_n^p|\mathcal{F}_n\right\} \overset{a.s.}{=} -a_n\left[\nabla_p L_{\mu_n,\delta_n}(p_n,\widetilde{u}_n) + o_\omega(\frac{1}{\sqrt{n}})\right] - b_n e^N \qquad (4.95)$$

Here the term $o_\omega(\frac{1}{\sqrt{n}})$ has appeared as a result of the application of the strong law of large numbers [28],[30]:

$$V_n - V \overset{a.s.}{=} o_\omega(\frac{1}{\sqrt{n}})$$

Also we have

$$\mathbf{E}\left\{\overline{\zeta}_n + \widetilde{u}_n|\mathcal{F}_n\right\} \overset{a.s.}{=} Vp_n + (1+\delta_n)\widetilde{u}_n = \nabla_{\widetilde{u}} L_{\mu_n,\delta_n}(p_n,\widetilde{u}_n) \qquad (4.96)$$

Taking into account the strictly convex property (4.72), we conclude that

$$2\gamma_n\left(p - p_n^*\right)^T \mathbf{E}\left\{A_n^p|\mathcal{F}_n\right\} - 2\gamma_n^u\left(\widetilde{u}_n^* - \widetilde{u}_{n+1}^*\right)^T \mathbf{E}\left\{\overline{\zeta}_n + \widetilde{u}_n|\mathcal{F}_n\right\} \overset{a.s.}{=}$$

$$= -2\gamma_n a_n\left(p - p_n^*\right)^T \nabla_p L_{\mu_n,\delta_n}(p_n,\widetilde{u}_n) -$$

$$-2\gamma_n^u\left(\widetilde{u}_n^* - \widetilde{u}_{n+1}^*\right)^T \nabla_{\widetilde{u}} L_{\mu_n,\delta_n}(p_n,\widetilde{u}_n) + \gamma_n a_n o_\omega(\frac{1}{\sqrt{n}}) \leq$$

$$\leq -2\gamma_n a_n \lambda_{\min}(H)\left(\left\|p_n - p_n^*\right\|^2 + \left\|\widetilde{u}_n - \widetilde{u}_n^*\right\|^2\right) + \gamma_n a_n o_\omega(\frac{1}{\sqrt{n}}) \qquad (4.97)$$

From (4.71) we derive

$$H = \begin{bmatrix} \delta I_N + V^T V & V^T \\ V & (1+\delta) I_m \end{bmatrix} = \begin{bmatrix} V^T V & V^T \\ V & I_m \end{bmatrix} + \begin{bmatrix} \delta I_N & 0 \\ 0 & \delta I_m \end{bmatrix} \geq$$

$$\geq \begin{bmatrix} \delta I_N & 0 \\ 0 & \delta I_m \end{bmatrix} = \delta I_{N+m}$$

and hence,

$$\lambda_{\min}(H) \geq \delta \qquad (4.98)$$

Substituting (4.98) into (4.97) we derive

$$2\gamma_n\left(p - p_n^*\right)^T \mathbf{E}\left\{A_n^p|\mathcal{F}_n\right\} - 2\gamma_n^u\left(\widetilde{u}_n^* - \widetilde{u}_{n+1}^*\right)^T \mathbf{E}\left\{\overline{\zeta}_n + \widetilde{u}_n|\mathcal{F}_n\right\} \overset{a.s.}{\leq}$$

a.s.

$$\leq -2\gamma_n a_n \delta_n\left(\left\|p_n - p_n^*\right\|^2 + \left\|\widetilde{u}_n - \widetilde{u}_n^*\right\|^2\right) + \gamma_n a_n o_\omega(\frac{1}{\sqrt{n}}) \leq$$

$$\leq -2\gamma_n a_n \delta_n\left(\left\|p_n - p_n^*\right\|^2 + \frac{\gamma_{n-1}}{\gamma_{n-1}^u}\left\|\widetilde{u}_n - \widetilde{u}_n^*\right\|^2\right) + \gamma_n a_n o_\omega(\frac{1}{\sqrt{n}}) =$$

$$= -2\gamma_n a_n \delta_n W_n + \gamma_n a_n o_\omega(\frac{1}{\sqrt{n}}) \tag{4.99}$$

Using (4.99) into (4.93) and taking into account that

$$a_n = O\left(\frac{\tau_n}{\widetilde{u}_n^+}\right)$$

after the application of the operator of conditional mathematical expectation and using the following inequality

$$2v_n\sqrt{W_n} \le v_n W_n + v_n, \; v_n, W_n \ge 0$$

we obtain

$$E\{W_{n+1}|\mathcal{F}_n\} \overset{a.s.}{\le} \left[1 - 2\gamma_n\tau_n\delta_n\left(\widetilde{u}_n^+\right)^{-1}\left(1 - \frac{2\widetilde{u}_n^+}{\gamma_n\tau_n\delta_n}\left|\frac{\gamma_n}{\gamma_n^u} - \frac{\gamma_{n-1}}{\gamma_{n-1}^u}\right|\right) + \right. \tag{4.100}$$

$$\left. + Cv_n\right] W_n + \gamma_n\tau_n\left(\widetilde{u}_n^+\right)^{-1} o_\omega(\frac{1}{\sqrt{n}}) + \overline{s}_n + Cv_n$$

where

$$\overline{s}_n = 4\gamma_n^2 + 2\left(1 + \frac{\gamma_{n-1}}{\gamma_{n-1}^u}\right)\left[C_1^2\left|\mu_n - \mu_{n+1}\right|^2 + C_2^2\left|\delta_n - \delta_{n+1}\right|^2\right] +$$

$$+2\frac{\gamma_{n-1}}{\gamma_{n-1}^u}\left(C_1\left|\mu_n - \mu_{n+1}\right| + C_2\left|\delta_n - \delta_{n+1}\right|\right)\left(C_3 + \gamma_n^u\sqrt{\theta_n}\right) +$$

$$+\frac{\gamma_{n-1}}{\gamma_{n-1}^u}\left(\gamma_n^u\right)^2\theta_n +$$

$$+2\left|\frac{\gamma_n}{\gamma_n^u} - \frac{\gamma_{n-1}}{\gamma_{n-1}^u}\right|\left[\left(\gamma_n^u\right)^2\theta_n + 2\left[C_1^2\left|\mu_n - \mu_{n+1}\right|^2 + C_2^2\left|\delta_n - \delta_{n+1}\right|^2\right]\right]$$

Taking into account assumption 3) of this theorem we can rewrite (4.100) as follows:

$$E\{W_{n+1}|\mathcal{F}_n\} \overset{a.s.}{\le} \left[1 - 2\gamma_n\tau_n\delta_n\left(\widetilde{u}_n^+\right)^{-1}(1 - d + o(1)) + Cv_n\right] W_n +$$

$$+\gamma_n\tau_n\left(\widetilde{u}_n^+\right)^{-1} o_\omega(\frac{1}{\sqrt{n}}) + \overline{s}_n + Cv_n \tag{4.101}$$

From this inequality and Robbins-Siegmund theorem [32] (see Appendix B) under the assumptions of this theorem we obtain the convergence with probability one, i.e.,

$$W_n \overset{a.s.}{\to} 0$$

The mean squares convergence follows from (4.101) after applying the operator of mathematical expectation to both sides of this inequality and using lemma A5 given in [4]. Theorem is proved. ■

Corollary 3. *If the parameters of the optimization algorithm, involved in the assumptions of theorem 7, belong to the following class of parameters*

$$\gamma_n := \frac{\gamma_0}{n^\gamma},\ \delta_n := \frac{\delta_0}{n^\delta},\ \mu_n := \frac{\mu_0}{n^\mu} \tag{4.102}$$

$$\tau_n := \frac{1}{N-1+n^\tau \ln n},\ \tilde{u}_n^+ := Const + n^u \ln n,\ Const > 0$$

$$\gamma,\ \delta,\ \mu,\ \tau > 0,\ u \geq 0$$

then

1) the convergence with probability one is ensured, i.e.,

$$W_n \overset{a.s.}{\to} 0$$

for

$$\gamma+\delta+u+\tau = 1,\ \gamma+\tau+u > \frac{1}{2} \tag{4.103}$$

$$2\mu > u+\tau,\ 2\delta > u+\tau$$

$$\theta := \min\{2\gamma,\ 2(1+\mu)-u-\tau,\ 2(1+\delta)-u-\tau,$$

$$2\gamma-u+\tau,\ 1+\mu+\gamma-u,\ 1+\delta+\gamma-u\} > 1$$

2) the mean squares convergence is guaranteed, i.e.,

$$\mathbf{E}\{W_n\} \to 0$$

for

$$\gamma+\delta+u+\tau = 1,\ 2\delta < 1,\ \theta > 1,\ 2\mu > u+\tau,\ 2\delta > u+\tau \tag{4.104}$$

Proof.

Notice that

$$\bar{s}_n = O(\frac{1}{n^\theta})$$

From the conditions of theorem 7 and (4.102) follows the desired result. ■

The next theorem gives the estimation of the order of convergence rate of this optimization algorithm.

Theorem 9. *Under the conditions of the previous theorem and for the class of parameters (4.102) there exist $\nu > 0$ such that*

$$W_n \overset{a.s.}{=} o_\omega\left(\frac{1}{n^\nu}\right),\ \mathbf{E}\{W_n\} = o\left(\frac{1}{n^\nu}\right) \tag{4.105}$$

where the order ν of convergence rate satisfies the following upper estimation

$$\nu < \nu^*(\gamma, \delta, \mu) \leq \nu^{**} = \frac{1}{3}$$

and the maximum convergence rate ν^{**} is reached for

$$\gamma = \gamma^* = \frac{2}{3}, \ \delta = \delta^* = \mu = \mu^* = \frac{1}{6}, \ \tau = \tau^* = u = u^* = \frac{1}{2} \quad (4.106)$$

$$\nu^*(\gamma^*, \delta^*, \mu^*) = \nu^{**}$$

Proof.

Notice that

$$W_n^* = \|p_n^* - p^{**}\|^2 + \|u_n^* - u^{**}\|^2 \leq$$
$$\leq 2W_n + C(\mu_n^2 + \delta_n^2)$$

Multiplting both sides of the last inequality by ν_n, we obtain

$$\nu_n W_n^* \leq 2\nu_n W_n + C\nu_n(\mu_n^2 + \delta_n^2)$$

In view of lemma A.3-2 (see appendix A) for $\nu_n := n^\nu$ we derive that the order must satisfies the following inequalities:

$$\nu < \nu^* = \min\left\{\theta - 1, \gamma + \tau + u - \frac{1}{2}, 1 + 2\mu - u - \tau, 1 + 2\delta - u - \tau, 2\mu, 2\delta\right\}$$

where the design parameters satisfy (4.8):

$$\gamma + \delta + u + \tau = 1, \ \gamma + \tau + u > \frac{1}{2}$$

$$2\mu > u + \tau, \ 2\delta > u + \tau, \ \theta > 1$$

and, similarly to the Lagrange multipliers case,

$$\delta \leq \tau$$

Maximization of ν^* with respect to the design parameters under the previous constraints, leads to

$$\mu = \delta, \ \frac{1}{2} - \delta = 2\delta, \ \theta - 1 = 1 - \delta$$

or

$$\delta = \delta^* = \mu = \mu^* = \frac{1}{6}, \ \theta = \theta^* = \frac{1}{3}, \ \nu^* = \frac{1}{3}$$

Taking into account that

$$u + \tau < 1$$

we conclude that

$$\begin{aligned}\theta^* &= \min\left\{2\gamma, \frac{7}{3}-(\frac{5}{6}-\gamma), 2\gamma-u+\tau, \frac{7}{6}+\gamma-u\right\} = \\ &= \min\left\{2\gamma, 2\gamma-u+\tau, \frac{7}{6}+\gamma-u\right\} = \frac{4}{3}\end{aligned}$$

if the corresponding parameters are respectively equal to

$$\gamma = \gamma^* = \frac{2}{3}, \ \tau = \tau^* = u = u^* = \frac{1}{2}$$

Theorem is proved. ■

Apart from the selection of some design parameters (γ_n, δ_n, etc.), the Lagrange multipliers and penalty function approaches generate different optimization algorithms.

In the next section we shall be concerned with the implementation aspects of these optimization algorithms.

4.9 Numerical examples

In this section computer simulations are presented to illustrate the performance of learning automata to solve constrained stochastic optimization problems on the basis of Lagrange multipliers and penalty function approaches.

Example 1 *In the first simulations, the following numerical example has been considered*

$$2p(1) + p(2) \to \min_{p \in S^2}$$

subject to

$$p(1) + 4p(2) \leq 1.5$$

This problem can be rewritten as follows

$$F^T p \to \min_{p \geq 0}$$

subject to

$$Ap \leq b$$

where

$$F^T = [2\text{-}1], \quad p = \begin{bmatrix} p(1) \\ p(2) \end{bmatrix}, \quad A = \begin{bmatrix} 1 & 1 \\ 1 & 1 \end{bmatrix}, \quad b = \begin{bmatrix} 1 \\ 1.5 \end{bmatrix}$$

The first line of the matrix A corresponds to the simplex constraint

$$p(1) + p(2) = 1$$

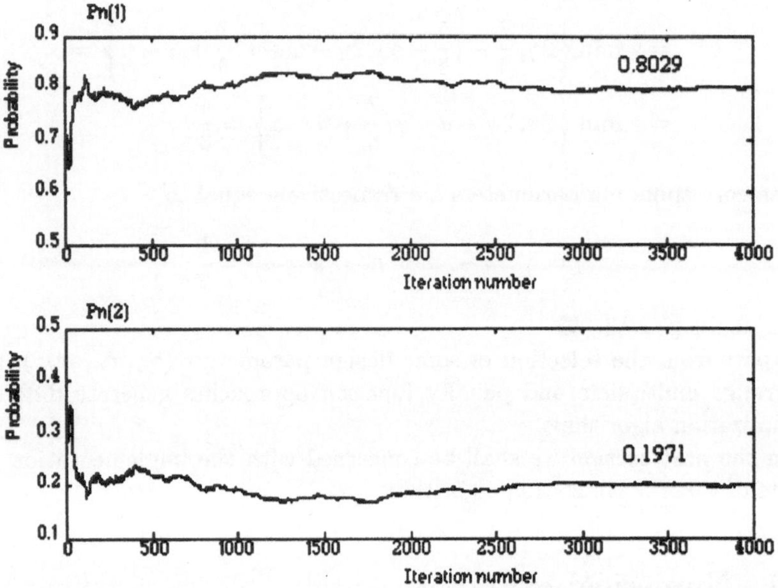

FIGURE 4.2. Evolution of the probabilities.

According to our notations, it follows:

$$v_1^0 = 2,\ v_2^0 = 1,\ v_1^1 = 1 - 1.5 = -0.5,\ v_2^2 = 4 - 1.5 = 2.5$$

The solution of this linear programming problem is

$$p(1) = \frac{5}{6}, \qquad p(2) = \frac{1}{6}$$

A two actions automaton has been considered. The automaton operates in a random environment which corresponds to the constrained stochastic optimization problem to be solved. This environment produces an output (response) different (continuous) from the reward/inaction (penalty) signal which gives partial information about their states.

Example 2 *The second set of simulations concerns the following numerical example:*

$$p(1) + 3p(2) + 5p(3) + p(4) + p(5) \to \min_{p \in S^5}$$

subject to

$$5p(1) + 2p(2) + p(3) + 5p(4) + 5p(5) \leq 2$$

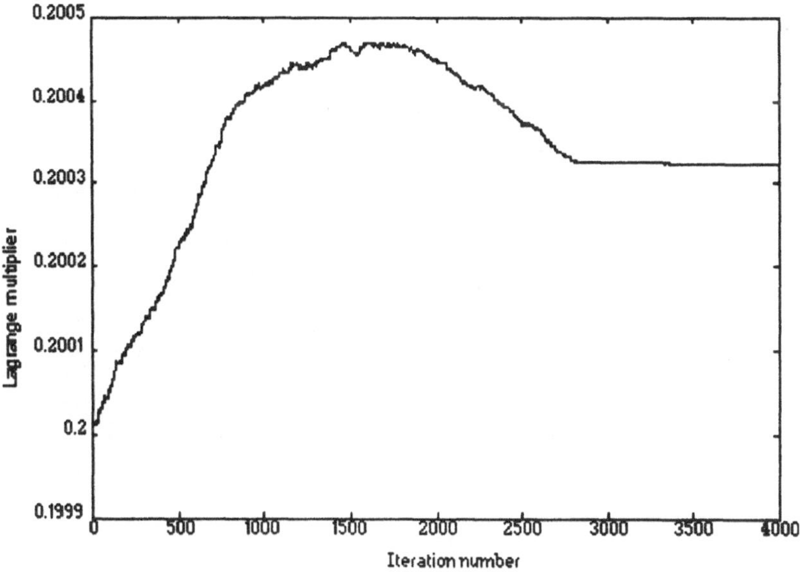

FIGURE 4.3. Evolution of the Lagrange multiplier.

$$p(1) + 4p(2) + 2p(3) + p(4) + p(5) \leq 3$$

This problem can be rewritten as follows

$$F^T p \rightarrow \min_{p \geq 0}$$

subject to

$$Ap \leq b$$

where

$$F^T = [2\text{-}3\ 5\ 1\ 1], \quad p = \begin{bmatrix} p(1) \\ p(2) \\ p(3) \\ p(4) \\ p(5) \end{bmatrix},$$

$$A = \begin{bmatrix} 1 & 1 & 1 & 1 & 1 \\ 5 & 2 & 1 & 5 & 5 \\ 1 & 4 & 2 & 1 & 1 \end{bmatrix}, \quad b = \begin{bmatrix} 1 \\ 2 \\ 3 \end{bmatrix}$$

The first line of the matrix A corresponds to the simplex constraint

$$p(1) + p(2)p(3) + p(4) + p(5) = 1$$

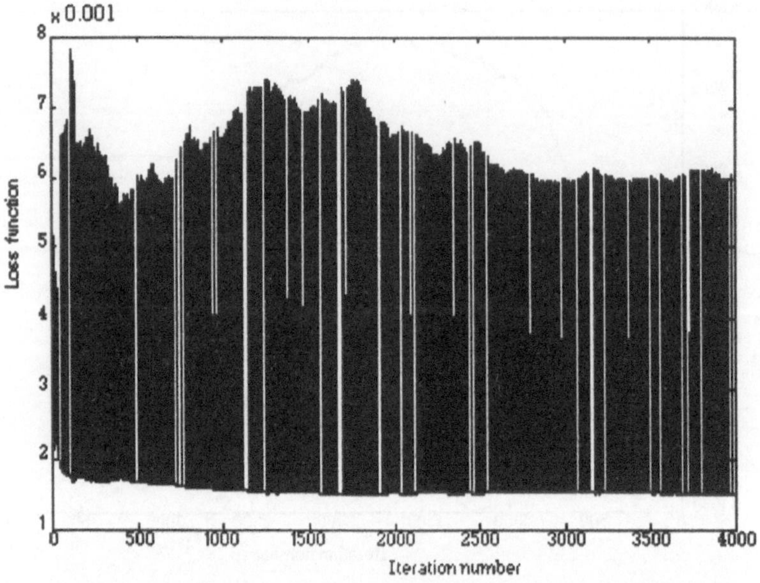

FIGURE 4.4. Evolution of the loss function.

According to our notations, it follows:

$$v_1^0 = 1, \ v_2^0 = 3, \ v_3^0 = 5, \ v_4^0 = 1, v_5^0 = 1$$

$$\begin{aligned} v_1^1 &= 5 - 2 = 3, \ v_2^1 = 2 - 2 = 0, \ v_3^1 = 1 - 2 = -1 \\ v_4^1 &= 5 - 2 = 3, \ v_5^1 = 5 - 2 = 3 \end{aligned}$$

$$\begin{aligned} v_1^2 &= 1 - 3 = -2, \ v_2^2 = 4 - 3 = 1, \ v_3^2 = 2 - 3 = -1 \\ v_4^2 &= 1 - 3 = -2, \ v_5^2 = 1 - 3 = -2 \end{aligned}$$

Notice that this constrained optimization problem has many solutions. For example:

$$p(1) = 0.0370, p(2) = 0.5566, p(3) = 0.3333, p(4) = 0.0370, p(5) = 0.0370$$

$$p(1) = 0.1111, p(2) = 0.5566, p(3) = 0.3333, p(4) = 0.0000, p(5) = 0.0000$$

$$p(1) = 0.0000, p(2) = 0.5566, p(3) = 0.3333, p(4) = 0.1111, p(5) = 0.0000$$

$$p(1) = 0.0000, p(2) = 0.5566, p(3) = 0.3333, p(4) = 0.0000, p(5) = 0.1111$$

An automaton of five actions has been considered.

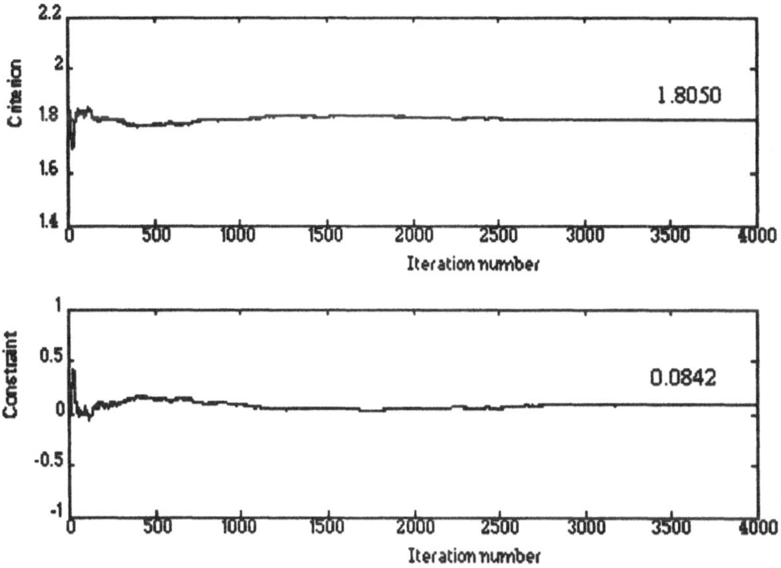

FIGURE 4.5. Evolution of the criterion and the constraint.

4.9.1 Lagrange Multipliers Approach

For every time n, the optimization algorithm based on the Lagrange multipliers approach performs the following steps:

- *Step 0.* Choose the parameters setting of the adaptation algorithms (4.33) and (4.34), the number of actions N and, initialize the probability vector p_1 $\left(p_1(i) = \frac{1}{N}, i = 1, ..., N\right)$ in the case where no prior information is available) and the other design parameters for example as follows

$$\lambda_1(j) = 1, \ \lambda_1 = 1, \ \sigma_j^+ = \frac{1}{2} + \max_i \left| v_i^j \right|, \ (j = 1, ..., m)$$

$$\delta_1 = \frac{1}{N+1} = \delta_0, \ \lambda_n^+ = 10^4 + \ln n$$

- *Step 1.* generate a uniformly distributed random variable, say z ($z \in (0,1)$) and select the control action $u(i)$ according to the following rule: the index i is given by

$$i \ = min \ j \text{ subject to } \sum_{j=1}^{i} p_n(j) \geq z$$

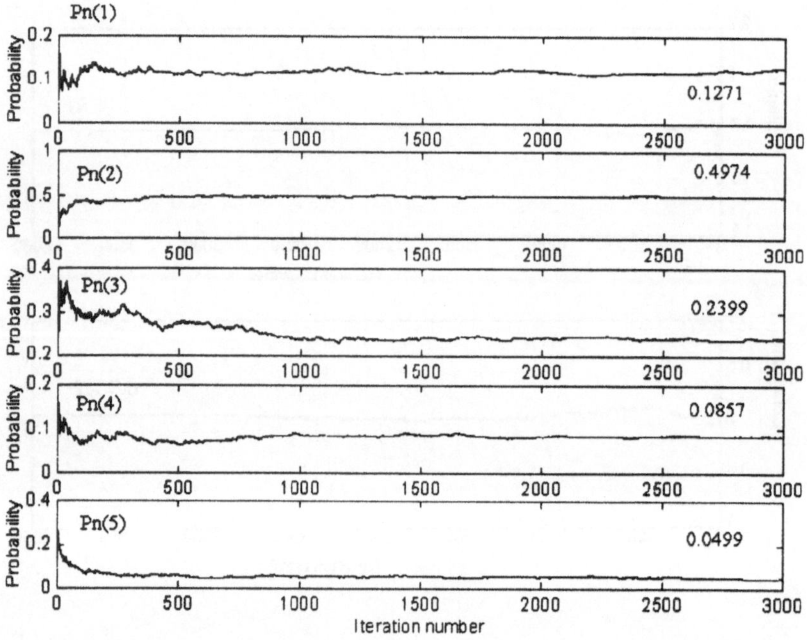

FIGURE 4.6. Evolution of the probabilities.

- *Step 2.* generate $(m+1)$ uniformly distributed random variable, say η_j $(\eta_j \in (0,1)\,,\ (j=0,...,m))$ and construct the vector ζ_n as follows

$$\zeta_n^j = \eta_j - \frac{1}{2} + v_i^j$$

- *Step 3.* construct the normalized automaton input ξ_n (environment response) according to (4.30) and (4.31).

- *Step 4.* adjust the probability distribution p_n and the Lagrange multipliers $\lambda_n(j)$ according to (4.33) and (4.34).

- *Step 5.* go to step 1 at the next time $n+1$.

The parameter setting for the Lagrange multipliers optimization algorithm are the following:

$$\lambda_1 = 0.2,\ \gamma_0 = 0.05,\ \gamma = 0.66,\ \delta_0 = 0.01,\ \tau = \delta = 0.16$$

Some simulations results related to example 1 are presented in the following.

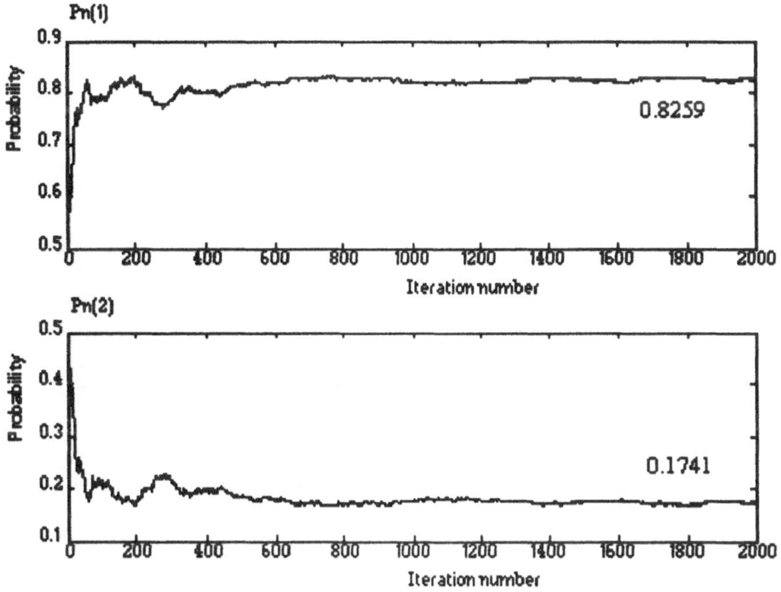

FIGURE 4.7. Evolution of the probabilities.

The variation of the components of the probability vector are depicted in Figure 4.2. These components converge respectively to $p_n(1) = 0.8029$ and $p_n(2) = 0.1971$.

Figure 4.3 presents the evolution of the Lagrange multiplier $\lambda_n(1)$ which converges after 2600 iterations. Figure 4.4 shows the loss function (ξ_n) variation. The evolution of, respectively the criterion (Φ_n^0) and the constraint (Φ_n^1) are given in Figure 4.5. The minimal value of the criterion is equal to 1.8050.

The following simulation results concern the second example. The evolution of the components of the probability vector $p_n(i) (i = 1, ..., 5)$ is represented in Figure 4.6.

The constraints are satisfied.drawn in Figure 4.9. The minimal value of the criterion is equal to 2.9704. The learning automaton provides a good optimization performance in the corresponding stochastic environment.

4.9.2 Penalty function approach

We will now attempt to solve the previous constrained stochastic optimization problems using the penalty function approach. The mechanization of the optimization algorithm based on the penalty function is similar to the previous one. The parameter setting for the optimization algorithm based

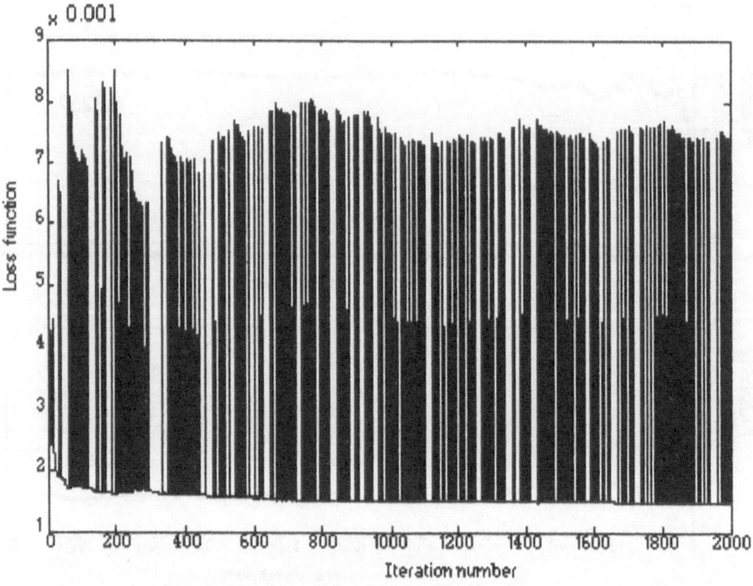

FIGURE 4.8. Evolution of the loss function.

on the penalty function approach are the following:

$$Const = 20, \ \gamma_0 = 0.05, \ \gamma = \frac{2}{3}, \ \delta_0 = 0.01, \ \mu_0 = 0.05$$

$$\mu = \delta = \frac{1}{6}, \ u = \tau = \frac{1}{2}$$

A learning automaton operating in a multi-teacher environment (with continuous response) has been implemented to solve the previous optimization problems on the basis of penalty function approach as described in the previous section.

For every time n, the optimization algorithm based on the penalty function approach performs a set of steps similar to the previous one (Lagrange approach).

For the first example (automaton with two actions and one constraints), the evolution of the probabilities is depicted in Figure 4.7. The evolution of the loss function is depicted in Figure 4.8. The evolution of, respectively the criterion (Φ_n^0) and the constraint (Φ_n^1) are given in Figure 4.9.

The criterion converges also to the same value 1.8244. These Figures indicate clearly the performance of learning automata when used for solving stochastic constrained optimization problems.

For the second example, the evolution of the probabilities is shown in Figure 4.10.

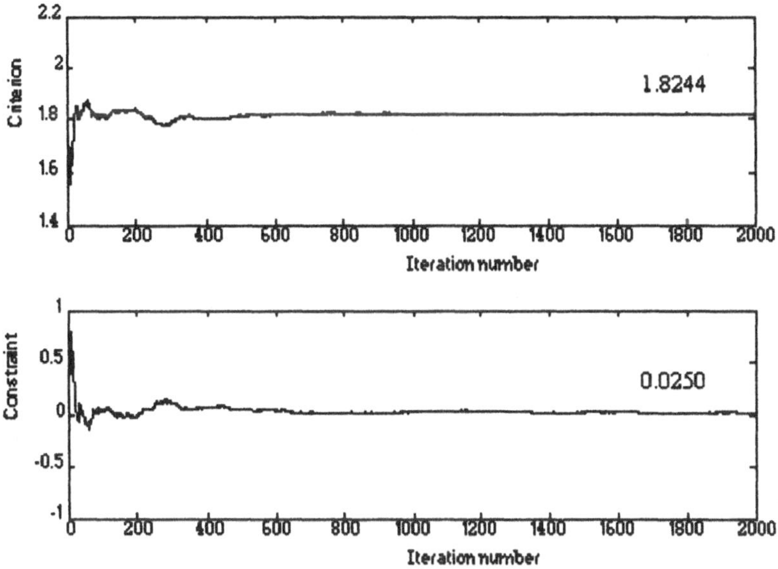

FIGURE 4.9. Evolution of the criterion and the constraint.

The constraint (Φ_n^1) is satisfied5. The criterion converges to 3.1207. It can be seen that:

the learning automaton has efficient learning ability

the rate of convergence is good.

The Lagrange multipliers and penalty function approaches lead to the same results. The above simulation results imply that learning automata can be used to solve constrained optimization problems. They can carry out a powerful search for the optimal solution and do not involve problem such as functional knowledge and computing robustness.

The authors would like thank Mr. Eduardo Gomez Ramirez for his assistance in carrying out these simulation results.

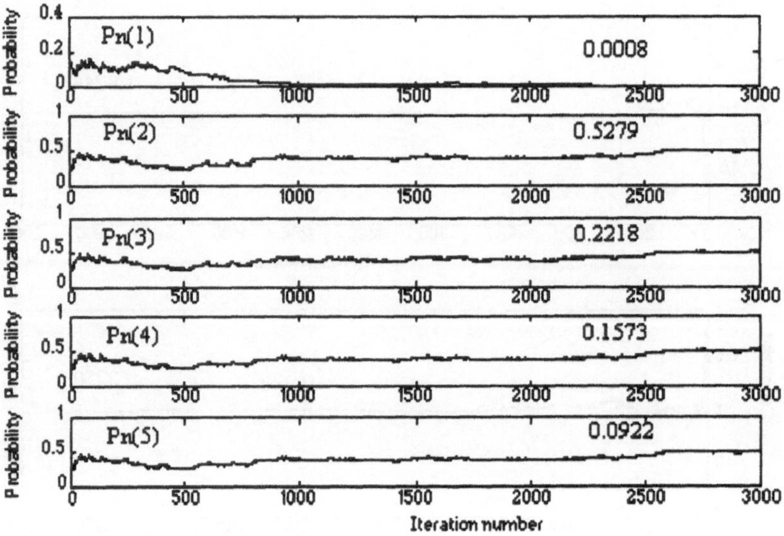

FIGURE 4.10. Evolution of the probabilities.

4.10 Conclusion

In this chapter we have formulated and solved a stochastic constrained optimization problem on finite set using a variable structure learning automaton operating in a random multi-teacher environment (construction of the environment response on the basis of the available measurements). The equivalence of this problem with the stochastic linear programming problem has been stated. Two approaches have been considered: Lagrange multipliers and penalty function. Two new tools: regularization and normalization have been introduced. The asymptotic behaviour (convergence and convergence rate) of these optimization algorithms (learning systems) has been investigated using martingales and Lyapunov approach. It has been shown that the probabilities tend to the optimal strategy. This theoretical study clearly show the performance of the behaviour of learning automata as a tool for solving stochastic constrained optimization problems.

The Lagrange multipliers and the penalty function approaches exhibit the same convergence rate ($o_\omega(n^{-1/3})$). In the penalty function approach we have been concerned with the estimation of the matrix related to the constraints. It has been shown in [24] that the convergence rate ($o_\omega(n^{-2/5})$), associated with projectional gradient scheme, is greater than the convergence rate of the optimization algorithm based on the Lagrange mulipliers and the penalty function procedures, presented in this chapter. Several simulation results have been presented. These results illustrate the feasibility and the performance of the learning automata in connection with the Lagrange multipliers and the penalty function approaches, to solve constrained stochastic optimization problems.

It suffices to say, in conclusion, that learning automata form a cornerstone for solving several complex engineering problems.

The previous chapters of this book have been mainly concerned with the use of learning automata operating in stationary random environments to solve stochastic optimization problems. However, in most situations there are a number of nonstationary processes which have to be taken into consideration. Methods for extending, in some sense the approach based on learning automata to the optimization problems involving nonstationary processes are discussed in the following chapter.

References

[1] Clarke F H, Dem'yanov V F, Gianessi F 1989 *Nonsmooth Optimization and Related Topics.* Plenum Press, New York

[2] Najim K 1989 *Process Modeling and Control in Chemical Engineering.* Marcel Dekker, New York

[3] Najim K, Oppenheim G 1991 Learning systems: theory and applications. *IEE Proceedings Computer and Digital Techniques* 138:183-192

[4] Najim K, Poznyak A S 1994 *Learning Automata: Theory and Applications.* Pergamon Press, Oxford

[5] Wong E 1973 Recent progress in stochastic processes - a survey", *IEEE Trans. on Information Theory.* 19:262-275

[6] Bertsekas D P 1982 *Constrained Optimization and Lagrange Multiplier Methods.* Academic Press, New York

[7] Rockafellar R T 1993 Lagrange multipliers and optimality. *SIAM Review* 35:183-238

[8] Rockafellar R T 1970 *Convex Analysis.* Princeton University Press, Princeton

[9] Sposito V A 1975 *Linear and Nonlinear Programming.* The Iowa State University Press/AMES

[10] Martos B 1975 *Nonlinear Programming Theory and Methods.* North-Holland Publishing Company, Amsterdam

[11] Whittle P 1971 *Optimization under Constraints.* Wiley-Interscience, New York

[12] D.G. Luenberger D G 1965 *Introduction to Linear and Nonlinear Programming.* Addison-Wesley, London

[13] Avriel M 1976 *Nonlinear Programming: Analysis and Methods.* Prentice-Hall, Englewood Cliffs

[14] Liukkonen J R, Levine A 1994 On convergence of iterated random maps. *SIAM J. Control and Optimization* 32:1752-1762

[15] Bush R R, Mosteller F 1958 *Stochastic Models for Learning.* John Wiley & Sons, New York

[16] Narendra K S, Thathachar M A L 1989 *Learning Automata an Introduction.* Prentice-Hall, Englewood Cliffs

[17] Najim K, Poznyak A S 1996 Multimodal searching technique based on learning automata with continuous input and changing number of actions. *IEEE Trans. on Systems, Man, and Cybernetics* 26:666-673

[18] Poznyak A S 1973 Learning automata in stochastic programming problems. *Automation and Remote Control* 34:1608-1619

[19] Baba N 1984 *New Topics in Learning Automata Theory and Applications.* Springer-Verlag, Berlin

[20] Polyak B T 1987 *Introduction to Optimization.* Optimization Software, Publication Division, New York

[21] Kaplinskii A I, Propoi A I 1970 Stochastic approach to nonlinear programming problems. *Automation and Remote Control* 31:448-459

[22] Kaplinskii A I, Poznyak A S, Propoi A I 1971 Optimality conditions for certain stochastic programming problems. *Automation and Remote Control* 32:1210-1218

[23] Kaplinskii A I, Poznyak A S, Propoi A I 1971 Some methods for the solution of stochastic programming problems. *Automation and Remote Control* 32:1609-1616

[24] Nazin A V, Poznyak A S 1986 *Adaptive Choice of Variants. (in Russian)* Nauka, Moscow

[25] Vajda S 1972 *Probabilistic Programming.* Academic Press, New York

[26] Zangwill W I 1969 *Nonlinear Programming: A Unified Approach.* Prentice-Hall, Englewood Cliffs

[27] Garcia C B, Zangwill W I 1981 *Pathways to Solutions, Fixed Points, and Equilibria.* Prentice-Hall, Englewood Cliffs

[28] Doob J L 1953 *Stochastic Processes.* John Wiley & Sons, New York

[29] Charnes A, Cooper W W, Thompson G J 1964 Critical path analysis via chance constrained and stochastic programming. *Operations Res.* 12:460-470

[30] Ash R B 1972 *Real Analysis and Probability.* Academic Press, New York

[31] Arrow K J,Hurwics L, Uzawa H 1961 Constraint qualifications and maximization problems. *Naval Res. Log. Quart.* 8:175-191

[32] Robbins H, Siegmund D 1971 A convergence theorem for nonnegative almost supermartingales and some applications. In: Rustagi J S (ed) 1971 *Optimizing Methods in Statistics.* Academic Press, New York

[33] Albert A 1972 *Regression and the Moore-Penrose Pseudoinverse.* Academic Press, New York

[34] Slater M 1950 Lagrange multipliers revisted: a contribution to nonlinear programming. *Cowles Discussion Paper* 403

[35] Spingarn J E, Rockafellar R T 1979 The generic nature of optimal conditions in nonlinear programs. *Mathematics of Operation Research* 4:425-430

5

Optimization of Nonstationary Functions

5.1 Introduction

In most engineering problems it is assumed that the disturbances are stationary. Hence this assumption on real sequence is really only one of the properties needed in order for the results in the stochastic analysis to be true. In practice, many optimization and control problems involve nonstationary functions and/or nonstationary observation noise. For example:

- Planning and operation of power systems lead to the prediction of power load. The prediction is needed for a variety of times, ranging from years down to fractions of an hour (short-term and long-term prediction). The power load can be regarded as a nonstationary random process. It has a noticeable seasonal pattern and a periodic structure where the main period is one week. It is influenced by e.g. domestic and industrial needs and long or short term weather conditions [1].

- In process design problems (process flow diagram and major equipments list determination), the objective function consists of the sum of the annualized investment cost plus the sum of the operating costs [2]. While the choice of the above objective function appears reasonable, it fails to account for the fact that a large uncertainty may be involved in future operational elements, such as parameter drift (transfer coefficients, reaction constants, etc.), disturbances (internal and external), fluctuations of the value of the money, of the quality of the feedstreams, etc.. Most of these perturbations are nonstationary.

- In industrial operations the optimum conditions change as time passes. Any observations are obscured by disturbances, and it is not easy to distinguish between true changes in optimum conditions and the spurious fluctuations caused by random perturbations.

- Nonstationary data appear in systems identification [3]-[4].

In this chapter the use of learning automata operating in asymptotically nonstationary environment in order to solve this type of optimization problems will be discussed. The results derived here enlarge considerably the area of applications to which learning automata may be applied.

The optimization problems considered in this chapter are stated in the next section.

5.2 Optimization of Nonstationary Functions

In this section we describe the general type of optimization problem we shall interested in.

Let us consider the optimization of nonstationary functions defined on finite sets:

$$f_n(u_n) \to \inf_{u_n} , \; u_n \in U = (u(1), ..., u(N)) \tag{5.1}$$

We assume that only disturbed data y_n of the function $f_t(u)$ are available, i.e.,

$$y_n = y_n(u_n, \omega) = f_n(u_n) + \zeta_n \tag{5.2}$$

Let us assume that all nonstationary effects belong to the class of "asymptotically stationary in average" processes, i.e. they satisfy the following assumptions:

(H1) there exists

$$E\{y_n/u_n = u(i) \wedge \mathcal{F}_{n-1}\} = f_n(u(i)) + \mathbf{E}\{\zeta_n/u_n = u(i) \wedge \mathcal{F}_{n-1}\} = \overline{f}_n(i) \tag{5.3}$$

(H2)

$$\frac{1}{n}\sum_{t=1}^{n} \overline{f}_t(i) \overset{a.s.}{\to} \widetilde{f}(i) \tag{5.4}$$

The first assumption states some kind of restriction to the properties of the observation noise. For example, **(H1)** will be satisfied if at each time n the noise ζ_n has a bounded second moment, i.e.,

$$\mathbf{E}\{\zeta_n^2\} < \infty$$

This property holds, for example, for Gaussian noises and is not true for noises having Cauchy distribution.

The second property (assumption) represents some kind of "stationarity in average" and will be satisfied for optimized functions and noises which are stationary in average. The following examples illustrate these characteristics

Example 1 *Let us consider the function given by (see Figure 5.1)*

$$f_n(u) = (u - u^* \sin \omega n - c)^2 , \; \omega, c > 0 \tag{5.5}$$

For independent centered normal distributed noises, it follows:

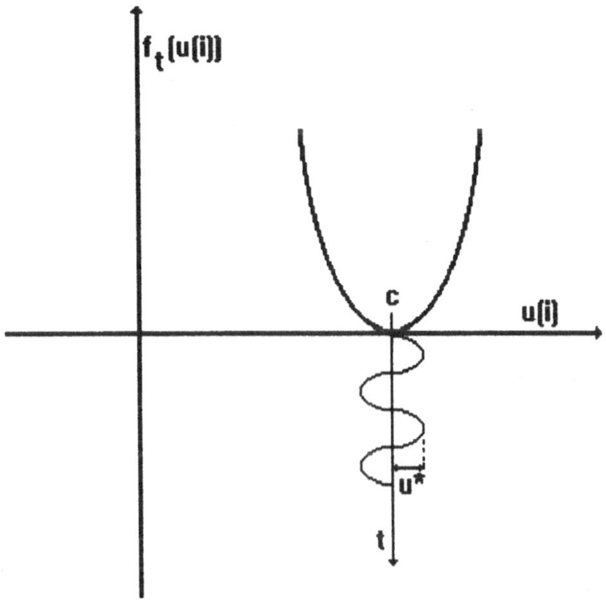

FIGURE 5.1. Nonstationary function.

$$\frac{1}{n}\sum_{t=1}^{n}\overline{f}_t(i) = \frac{1}{n}\sum_{t=1}^{n}\left[f_t(u(i)) + \mathbf{E}\{\zeta_t/u_t = u(i) \wedge \mathcal{F}_{t-1}\}\right] =$$
$$= \frac{1}{n}\sum_{t=1}^{n}(u(i) - u^*\sin\omega t - c)^2 = (u(i)-c)^2 -$$
$$-2\,(u(i)-c)\,u^*\frac{1}{n}\sum_{t=1}^{n}\sin\omega t + (u^*)^2\,\frac{1}{n}\sum_{t=1}^{n}\sin^2\omega t$$

*and hence, assumption **(H2)** is fulfilled*

$$\frac{1}{n}\sum_{t=1}^{n}\overline{f}_t(i) \underset{n\to\infty}{\to} (u(i)-c)^2 + \frac{(u^*)^2}{2} := \widetilde{f}(i)$$

Example 2 *Independent Gaussian noises $\mathcal{N}(a_t,\sigma^2)$ where a_t is periodically time-varying and satisfies the following constraint*

$$\frac{1}{n}\sum_{t=1}^{n}a_t \underset{n\to\infty}{\to} 0$$

For this example, the previous assumptions hold.

Throughout this chapter we will assume that these two assumptions ((**H1**)and(**H2**)) are satisfied. Under these assumptions we will be interested in the following stochastic optimization problem given on a finite set:

$$\overline{\lim_{n\to\infty}} \frac{1}{n}\sum_{t=1}^{n} y_n \to \inf_{\{u_t\}} \tag{5.6}$$

This stochastic optimization problem (5.6) presents the correct mathematical formulation of the nonstationary optimization problem (5.1), using the observations (5.2).

Hereafter we will associate these observations of the disturbed function to be optimized with the response of an environment response. Consequently, we can associate this stochastic optimization problem on finite sets (5.6) to the behaviour of a learning automaton operating in a random environment which is asymptotically stationary in average i.e.,

$$y_t = \xi_t \tag{5.7}$$

The next section presents an overview of nonstationary environments.

5.3 Nonstationary learning systems

Learning systems are an efficient tool to deal with a large number of engineering problems [5]-[6]-[7]-[8]. A learning system interacts with an environment and learns the optimal action which the environment offers. Most of the available studies relate to the behaviour of learning automata in stationary environments [5]-[8]. The problem concerning the behaviour of learning automata in nonstationary environments is difficult, and few results are known [7]-[9]-[10]-[11]-[12]-[13]. Narendra and Viswanathan [13] considered periodically changing nonstationary random medium with unknown period. A nonstationary environment in which one action u_α continues to have the minimum penalty c_α even though all the penalty probabilities keep changing with time, i.e., $c_\alpha(t,\omega)+ \delta < c_j(t,\omega)$ holds for some α, some $\delta > 0$, and for all j ($j \neq \alpha$), and for all random factors ω; has been studied by Baba and Sawaragi [14]. These results have been extended to the nonstationary multi-teacher environment [12]. Several basic norms (expediency, optimality, etc.) of the learning behaviour of stochastic automaton in multi-teacher environment are given in [12]. A variable-structure stochastic automaton where the penalty probabilities have been assumed to be time-varying and depending on the input action chosen, was introduced by Narendra and Thathachar [15]. Srikantakumar and Narendra [9] have developed an adaptive routine in telephone networks using learning methods. They have considered a nonstationary environment for which reward probabilities $c_i(p)$ and their derivatives are Lipschitz functions of all their arguments and

$$\frac{\partial c_i(p)}{\partial p_i} > 0 \ \forall i \text{ and } \frac{\partial c_j(p)}{\partial p_i} << \frac{\partial c_i(p)}{\partial p_i} \text{ for } j \neq i$$

So, this environment is nonstationary because the probability vector p changes with time $p = p(n)$ (the probability vector is updated using a reinforcement scheme).

A L_{R-P} scheme parametrized at step n by an intermediate parameter vector $\Theta_n \in R^n$ has been suggested by Barto et al.[10]. This scheme has been used to solve a class of learning tasks that combines aspects of learning automaton tasks and supervised learning pattern-classification tasks (associative reinforcement tasks). In [16], the environment (the penalty probabilities) was modelled by a difference equation for the prediction of the transient behaviour of the learning system. Another source of nonstationary environment is in the concept of multilevel systems of automata which has been introduced by Thathachar and Ramakrishan [17], and Najim and Poznyak [5]. A nonstationary environment arises indirectly in connection with hierarchical system of learning automata [7]. It has been shown in [5] that the use of hierarchical system of learning automata accelerates the learning process. The latter reference also discusses in fair detail some of the applications of hierarchical structure of learning automata. Baba and Mogami [11] have shown that an extended form of the scheme proposed by Thathachar and Ramakrishan [17] ensures absolute expediency in a nonstationary environment having the property that there exists a unique path which receives the least sum of the penalty strengths in the sense of mathematical expectation.

In this work, we consider the behaviour of learning automata operating in a nonstationary environment. The conditional expectation of the environment responses are assumed to be time-varying. A normalization procedure is introduced to deal with environment responses which do not belong to the segment $[0, 1]$. The class of asymptotically stationary environment (in the average sense) is also introduced. We show that several nonstationary environments such as:

- Markovian environments
- Periodically changing environments
- Multi-teacher environments
- Environments with fixed optimal actions

  belong to this class of asymptotically stationary environments.

Several theoretical results are stated. These results concern the properties of reinforcement schemes, normalized environment response and the asymptotically optimal behaviour of different learning automata.

These results will be divided into a succession of lemmas and theorems for greater clarity. Some key lemmas are given in Appendix A.

The learning automaton and the nonstationary environment are described in the next section.

5.4 Learning Automata and Random Environments

In this section we consider the behaviour of a variable-structure stochastic automaton in a nonstationary environment. The interaction between the stochastic automaton and the environment is shown in Figure 5.2.

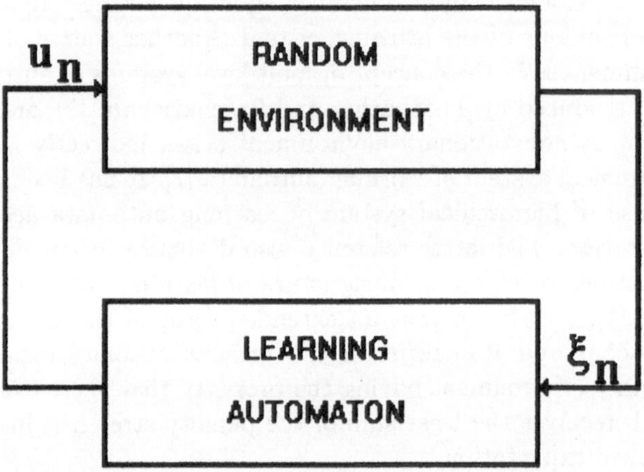

FIGURE 5.2. Learning automaton interacting with environment.

The role of the environment (medium) is to establish the relation between the actions of the automaton and the signals received at its input.

Let us recall the definition of a learning automaton. An automaton is a sequential machine described by (see the previous chapter):

$$\{\Xi, U, \mathcal{R}, \{\xi_n\}, \{u_n\}, \{p_n\}, T\}$$

where:

(i) Ξ is the automaton input bounded set.

(ii) U denotes the set $\{u(1), u(2), ..., u(N)\}$ of actions of the automaton.

(iii) $\mathcal{R} = (\ \Omega, \mathcal{F}, \mathbf{P}\ )$ a probability space.

(iv) $\{\xi_n\}$ is a sequence of automaton inputs (environment response, $\xi_n \in \Xi$) provided by the environment in a binary (P-model environment) or continuous (S-model environment) form.

(v) $\{u_n\}$ is a sequence of automaton outputs (actions).

(vi) $p_n = [p_n(1), p_n(2), ..., p_n(N)]^T$ is the probability distribution at time n

$$p_n(i) = \mathbf{P}\{\omega : u_n = u(i) \ / \ \mathcal{F}_{n-1}\} \text{ and } \sum_{i=1}^{N} p_n(i) = 1 \ ,\forall n$$

where $\mathcal{F}_n = \sigma(\xi_1, u_1, p_1; ...; \xi_n, u_n, p_n)$ is the σ-algebra generated by the corresponding events $(\mathcal{F}_n \in \mathcal{F})$.

(vii) $c_n = [c_n(1), c_n(2), ..., c_n(N)]^T$ is the conditional mathematical expectation vector of the environment responses (at time n).

(viii) T represents the reinforcement scheme (updating scheme) which changes the probability vector p_n to p_{n+1}:

$$p_{n+1} = p_n + \gamma_n T_n(p_n; \{\xi_t\}_{t=1,...,n}; \{u_t\}_{t=1,...,n}) \tag{5.8}$$

$$p_1(i) > 0 \quad \forall i = 1, ..., N$$

where γ_n is a scalar correction factor and the vector $T_n(.) = \left[T_n^1(.), ..., T_n^N(.)\right]^T$ satisfies the following conditions (for preserving probability measure):

$$\sum_{i=1}^{N} T_n^i(.) = 0, \ \forall n \tag{5.9}$$

$$p_n(i) + \gamma_n T_n^i(.) \in [0, 1] \qquad \forall n, \forall i = 1, ..., N \tag{5.10}$$

This is the heart of the learning automaton. Different reinforcement schemes satisfying these conditions (5.9),(5.10) are given in Table 2.1 [5].

In this study we shall be concerned with Bush-Mosteller [18], Shapiro-Narendra [19] and Varshavskii-Vorontsova [20] reinforcement schemes.

As in the previous chapters, the loss function Φ_n associated with the learning automaton is given by

$$\Phi_n = \frac{1}{n} \sum_{t=1}^{n} \xi_t \tag{5.11}$$

Let us consider the nonstationary environment which is characterized by the following two properties:

(H3) The conditional mathematical expectations of the environment responses exists, i.e.,

$$\mathbf{E}\left\{\xi_n / \mathcal{F}_{n-1} \bigwedge u_n = u(i)\right\} = c_n(i), \quad \forall \ i = 1, ..., N \tag{5.12}$$

and their arithmetic averages tend to a finite limit, with probability one i.e.,

$$\frac{1}{n}\sum_{t=1}^{n} c_t(i) \underset{n\to\infty}{\overset{a.s.}{\to}} \widetilde{c}(i) \tag{5.13}$$

(H4) The conditional variances of the environment responses are uniformly bounded, i.e.,

$$\begin{gathered} \mathbf{E}\left\{(\xi_n - c_n(i))^2 / \mathcal{F}_{n-1} \bigwedge u_n = u(i)\right\} = \sigma_n^2(i) \quad \forall\, i = 1, ..., N \\ \max_i \sup_n \sigma_n^2(i) = \sigma^2 < \infty \end{gathered} \tag{5.14}$$

Definition 3 *A random environment, satisfying conditions (5.12)-(5.14), will be said* ***"asymptotically stationary in the average sense"*** *and will be denoted by $\mathcal{K}$.*

The following nonstationary environments, which have been considered by several authors, are asymptotically stationary in the average sense.

- *Markovian environment* [6].

  A set of M stationary environments which conditional mathematical expectations of the environment responses are equal to $c^j(i)$ $(j = 1, ..., M)$ is considered. According to a stationary transition matrix $\pi = \|\pi_{ij}\|_{i,j=M}$ of an ergodic Markovian chain, the automaton switches from one environment to another environment. In this case, we have:

$$\mathbf{E}\left\{\xi_n / \mathcal{F}_{n-1} \bigwedge u_n = u(i)\right\} = \sum_{j=1}^{M} c^j(i)\; r_n(j) \quad \forall\, i = 1, ..., N$$

$$r_n(j) = \sum_{s=1}^{M} \pi_{sj}\; r_n(s) \quad \underset{n\to\infty}{\to} r(j) \quad \forall\, j = 1, ..., M$$

$$\frac{1}{n}\sum_{t=1}^{n} \mathbf{E}\left\{\xi_t / \mathcal{F}_{t-1} \bigwedge u_t = u(i)\right\} \underset{n\to\infty}{\to} \widetilde{c}(i) = \sum_{j=1}^{M} c^j(i)\; r(j)$$

$$\forall\, i = 1, ..., N$$

- *Periodically changing nonstationary environments* [21]-[13]-[7].

  For the class of sinusoidally varying environment we have:

$$c_t(i) = c^0(i) + a(i) \sin\left(\frac{2\pi}{T} n\right), \quad \frac{1}{n}\sum_{t=1}^{n} c_t(i) \underset{n\to\infty}{\overset{a.s.}{\to}} \widetilde{c}(i) = c^0(i)$$

- *Multi-teacher environments* [11]-[12].

  In this case, the input of the learning automata corresponds to the arithmetical average of the continuous (or binary) outputs ξ_n^i $(i = 1, ..., M)$ of the M stationary environments ("teachers") responses, i.e.,

  $$\xi_n = \frac{1}{M}(\xi_n^1 + ... + \xi_n^M)$$

  Hence,

  $$\mathbf{E}\left\{\xi_n/\mathcal{F}_{n-1} \bigwedge u_n = u(i)\right\} = \frac{1}{M}\sum_{j=1}^{M} c^j(i) = c_t(i) \quad \forall\, i = 1, ..., N$$

  $$c^j(i) := \mathbf{E}\left\{\xi_n^j/\mathcal{F}_{n-1} \bigwedge u_n = u(i)\right\}$$

  $$\frac{1}{n}\sum_{t=1}^{n} c_t(i) = \widetilde{c}(i) = \frac{1}{M}\sum_{j=1}^{M} c^j(i) \quad \forall\, i = 1, ... N$$

- *Environments with fixed optimal actions* [14]-[5].

  In this case, it is assumed that there exists an action α which remains to be optimal even though the characteristics of the environment keep changing with time:

  $$\limsup_{n \to \infty} \mathbf{E}\left\{\xi_n/\mathcal{F}_{n-1} \bigwedge u_n = u(\alpha)\right\} <$$

  $$< \liminf_{n \to \infty} \mathbf{E}\left\{\xi_n/\mathcal{F}_{n-1} \bigwedge u_n = u(i)\right\}, \quad i \neq \alpha$$

The problem to be considered in this study is presented in the next section.

5.5 Problem statement

In the ideal situation (the conditional mathematical expectations of the environment responses are a priori known) we can easily define the optimal strategy $\{u_n^*\}$ as follows:

$$\begin{aligned} u_n^* &= u(\alpha_n)\ \forall\, n = 1, 2... \\ \alpha_n &= \arg\min_i c_n(i) \end{aligned} \tag{5.15}$$

The probability vector sequence $\{p_n^*\}$ corresponding to this optimal strategy (5.15) is equal to:

$$p_n^*(\alpha_n) = 1,\ p_n^*(j) = 0,\ j \neq \alpha_n,\ \forall\, n = 1, 2... \tag{5.16}$$

The automaton outputs (environment responses) and the loss function corresponding to this optimal strategy will be respectively denoted by $\{\xi_n^*\}$ and Φ_n^*.

In this chapter we shall be concerned with the following problem.

Problem statement:

For the class $\mathcal{K}$ (5.12)-(5.14) of environments and the reinforcement scheme (5.8), estimate the maximum asymptotic deviation J of Φ_n from Φ_n^*, i.e.,

$$J = J(\{p_n\}) = \sup_{\mathcal{K}} \limsup_{n\to\infty} [\Phi_n - \Phi_n^*] \tag{5.17}$$

5.6 Some Properties of the Reinforcement Schemes

In this section, the asymptotic deviation J (5.17) is expressed as a function of the penalty probabilities $\{c_n\}$ (5.12) and the probability vector sequence $\{p_n\}$ (5.8) and it is shown that under assumption **(H3)**, **(H4)** and for a class of correction factors, this deviation tends to zero, with probability one. The following lemmas state these results.

Lemma 1. *Assume that (H3) and (H4) hold, then the asymptotic deviation J is given by:*

$$J \overset{a.s.}{=} \sup_{\mathcal{K}} \limsup_{n\to\infty} \frac{1}{n}\sum_{t=1}^{n}\sum_{i=1}^{N} [c_t(i) - c_t(\alpha_t)]\, p_t(i) \tag{5.18}$$

Proof.

Let us introduce the following sequence

$$\delta_n := \frac{1}{n}\sum_{t=1}^{n}(\xi_t - \xi_t^*) - \frac{1}{n}\sum_{t=1}^{n}\sum_{i=1}^{N}[c_t(i) - c_t(\alpha_t)]\, p_t(i) \tag{5.19}$$

As $n \to \infty$, this sequence tends $(a.s.)$ to zero. Indeed, according to **(H3)**, it follows

$$\frac{1}{n}\sum_{t=1}^{n}\sum_{i=1}^{N}[c_t(i) - c_t(\alpha_t)]\, p_t(i) \overset{a.s.}{=} \frac{1}{n}\sum_{t=1}^{n}\mathbf{E}\{(\xi_t - \xi_t^*)/\mathcal{F}_{t-1}\} \tag{5.20}$$

Then, using the previous equality (5.20) we derive the recurrent form of (5.19):

$$\delta_n = \left(1 - \frac{1}{n}\right)\delta_{n-1} + \frac{1}{n}\left[(\xi_n - \xi_n^*) - \mathbf{E}\left\{(\xi_n - \xi_n^*)/\mathcal{F}_{n-1}\right\}\right] \tag{5.21}$$

Taking the conditional expectation of δ_n^2, and in view of **(H4)**, we derive the following inequality:

$$\mathbf{E}\left\{\delta_n^2/\mathcal{F}_{n-1}\right\} = \left(1 - \frac{1}{n}\right)^2 \delta_{n-1}^2 +$$

$$+\frac{2}{n}\left(1 - \frac{1}{n}\right)\mathbf{E}\left\{\delta_n\left[(\xi_n - \xi_n^*) - \mathbf{E}(\xi_n - \xi_n^*)/\mathcal{F}_{n-1}\right]/\mathcal{F}_{n-1}\right\} +$$

$$+\frac{1}{n^2}\mathbf{E}\left[(\xi_n - \xi_n^*) - \mathbf{E}\left\{(\xi_n - \xi_n^*)/\mathcal{F}_{n-1}\right\}\right]^2/\mathcal{F}_{n-1}\}$$

$$= \left(1 - \frac{1}{n}\right)^2 \delta_{n-1}^2 +$$

$$+\frac{1}{n^2}\mathbf{E}\left\{\left[(\xi_n - \xi_n^*) - \mathbf{E}\left\{(\xi_n - \xi_n^*)/\mathcal{F}_{n-1}\right\}\right]^2/\mathcal{F}_{n-1}\right\}$$

$$\overset{a.s.}{\leq} \left(1 - \frac{1}{n}\right)^2 \delta_{n-1}^2 + \frac{2}{n^2}\sigma^2$$

In view of lemma A. 11 [5], this inequality leads to

$$\delta_n \underset{n\to\infty}{\overset{a.s.}{\to}} 0$$

This fact implies the desired result (5.18). ■

Lemma 2. *Assume that assumptions (H3) and (H4) hold, and suppose that*

$$\gamma_n = \frac{\gamma}{n+a} \quad (\gamma\ ,\ a > 0) \tag{5.22}$$

Then, the asymptotic deviation J can be expressed as follows:

$$J \overset{a.s.}{=} \sup_{\mathcal{K}}\ \limsup_{n\to\infty} \widetilde{\Delta}^T \left[p_n - \frac{1}{n}\sum_{t=1}^{n} T_t\right] \tag{5.23}$$

where the components $\widetilde{\Delta}(i)$ of the vector $\widetilde{\Delta} = [\widetilde{\Delta}(1), ..., \widetilde{\Delta}(N)\ ]^T$ are equal to

$$\widetilde{\Delta}(i) = \widetilde{c}(i) - \widetilde{c}(\alpha) \quad \forall i = 1, ..., N \tag{5.24}$$

$$\alpha = \arg\ \min_i \widetilde{c}(i)$$

Proof.

Let us introduce the following random variable J_n :

$$J_n = \frac{1}{n}\sum_{t=1}^{n}\sum_{i=1}^{N}\left[c_t(i) - c_t(\alpha_t)\right] p_t(i) \tag{5.25}$$

and, consider the matrix version of the Abel's identity:

$$\sum_{t=n_0}^{n} A_t B_t = A_n \sum_{t=n_0}^{n} B_t - \sum_{t=n_0}^{n}\left[A_t - A_{t-1}\right]\sum_{s=n_0}^{t-1} B_s \tag{5.26}$$

$$A_t \in R^{m \times k}, \quad B_t \in R^{k \times l}$$

The proof of this identity is given in Appendix A (lemma A.5-1 in Appendix A).

Based on the Abel's identity (5.26), the functional (5.25) leads to:

$$J_n = \frac{1}{n}\sum_{t=1}^{n} p_t^T \Delta_t = p_n^T\left[\frac{1}{n}\sum_{t=1}^{n}\Delta_t\right] - \frac{1}{n}\sum_{t=1}^{n}\left(p_t - p_{t-1}\right)^T\sum_{s=1}^{t-1}\Delta_s =$$

$$= p_n^T\widetilde{\Delta}_n - \frac{1}{n}\sum_{t=1}^{n}(t-1)\ \gamma_{t-1}T_{t-1}^T\widetilde{\Delta}_{t-1}$$

Let us introduce the following functions:

$$\Delta_t\ (i) := c_t(i) - c_t(\alpha_t), \quad \widetilde{\Delta}_n := \frac{1}{n}\sum_{t=1}^{n}\Delta_t \tag{5.27}$$

Taking into account assumptions **(H3)**, **(H4)** and (5.22) we conclude that

$$\widetilde{\Delta}_n \overset{a.s.}{=} \widetilde{\Delta} + o_\omega(1)$$

$$n\,\gamma_n = \gamma + o_\omega(1), \quad n \to \infty$$

and

$$J_n = p_n^T\left[\widetilde{\Delta} + o_\omega(1)\right] - \frac{1}{n}\sum_{t=n_0}^{n}\left[\gamma + o_\omega(1)\right]T_{t-1}^T\left[\widetilde{\Delta} + o_\omega(1)\right] = \tag{5.28}$$

$$= \widetilde{\Delta}^T p_n - \gamma\widetilde{\Delta}^T\left[\frac{1}{n}\sum_{t=n_0}^{n} T_{t-1}\right] + o_\omega(1)$$

where $o_\omega(1)$ is a random bounded sequence which tends to zero, with probability one.

As $n \to \infty$, equation (5.28) leads to the desired result (5.23). ■

Lemma 3. *Assume that assumptions (H3), (H4) and (5.22) hold. If the reinforcement scheme (5.8) possesses the following properties:*

- *(P1)*
$$p_n(\alpha) \underset{n\to\infty}{\overset{a.s.}{\to}} 1 \tag{5.29}$$
- *(P2)*
$$\frac{1}{n}\sum_{t=n_0}^{n} T_t \underset{n\to\infty}{\overset{a.s.}{\to}} 0$$

then, the asymptotic deviation J is equal to zero, with probability one, i.e.,

$$J \overset{a.s.}{=} 0 \tag{5.30}$$

or, in other words, the loss function Φ_n(5.11) tends to its minimal possible value

$$\Phi_n \underset{n\to\infty}{\overset{a.s.}{\to}} \widetilde{c}(\alpha) \tag{5.31}$$

Proof.

Using expression (5.28) and taking into account that $\widetilde{\Delta}(\alpha) = 0$, we derive (5.30). ■

In the following we shall be interested in the analysis of different reinforcement schemes which are supplied by normalized environment responses.

5.7 Reinforcement Schemes with Normalization Procedure

This section deals with the normalization procedure which is useful when the environment responses do not belongs to the segment [0, 1]. This normalization procedure has been initially suggested by Najim and Poznyak [5] to solve the optimization problems related to multimodal functions, and for neural networks synthesis. It is described by the following algorithm:

$$\widetilde{\xi}_n := \frac{\left[s_n(i) - \min\limits_j s_{n-1}(j)\right]_+}{\max\limits_k \left[s_n(k) - \min\limits_j s_{n-1}(j)\right]_+ + 1}, \qquad u_n = u(i) \tag{5.32}$$

where

$$s_n(i) := \frac{\sum\limits_{t=1}^{n} \xi_t \chi(u_t = u(i))}{\sum\limits_{t=1}^{n} \chi(u_t = u(i))}, \qquad i = 1, ..., N \tag{5.33}$$

$$[x]_+ := \begin{cases} x & if \quad x \geq 0 \\ 0 & if \quad x < 0 \end{cases}, \quad \chi(u_n = u(i)) := \begin{cases} 1 & if \quad u_n = u(i) \\ 0 & if \quad u_n \neq u(i) \end{cases} \tag{5.34}$$

The properties of the normalized environment response will be summarized in the following lemma.

Lemma 4. *Assume that assumptions (H3) and (H4) hold and suppose that the reinforcement scheme (5.8) generates the sequence $\{p_n\}$ such that for any $i = 1, ..., N$*

$$\sum_{n=1}^{\infty} p_n(i) \overset{a.s.}{=} \infty \tag{5.35}$$

Then, the normalized environment response $\tilde{\xi}_n$ (5.32) possesses the following properties:

- *The number of selections of each action is infinite, i.e.,*

$$\sum_{t=1}^{\infty} \chi(u_t = u(i)) \overset{a.s.}{=} \infty \quad \forall i = 1, ..., N \tag{5.36}$$

- *The random variable $s_n(i)$ is asymptotically equal to the arithmetic average of the conditional expectation of the corresponding environment responses, i.e.,*

$$s_n(i) = \frac{1}{n} \sum_{t=1}^{n} c_t(i) + o_\omega(1), \qquad \forall i = 1, ..., N \tag{5.37}$$

- *For the selected action $u_n = u(i)$ at time n, the normalized environment reaction is asymptotically equal to $\tilde{\Delta}(i)$, i.e.,*

$$\tilde{\xi}_n \overset{a.s.}{=} \tilde{\Delta}(i) + o_\omega(1), \quad u_n = u(i) \quad \forall i = 1, ..., N \tag{5.38}$$

- *For the optimal action $u_n = u(\alpha)$, the normalized environment reaction is asymptotically equal to 0, i.e.,*

$$\tilde{\xi}_n \overset{a.s.}{=} o_\omega(1), \quad u_n = u(\alpha) \tag{5.39}$$

Proof.

1. (5.36) follows directly from assumption (5.35) and the Borel-Cantelli lemma [22].

2. Let us introduce the following sequence

$$\theta_n(i) := s_n(i) - \frac{\sum_{t=1}^{n} c_t(i)\chi(u_t = u(i))}{\sum_{t=1}^{n} \chi(u_t = u(i))}, \quad i = 1, ..., N$$

which leads to the following recurrent form of $\theta_n(i)$

$$\theta_n(i) = (1 - \lambda_n(i))\theta_{n-1}(i) + \lambda_n(i)\left[\xi_n - c_n(i)\right]$$

where

$$\lambda_n(i) := \frac{\chi(u_n = u(i))}{\sum_{t=1}^{n} \chi(u_t = u(i))}$$

Taking into account that (see assumptions **(H3)**,**(H4)**)

$$\mathbf{E}\left\{\left[\xi_n - c_n(i)\right]/\mathcal{F}_{n-1} \bigwedge u_n = u(i)\right\} \overset{a.s.}{=} 0$$

it follows that

$$\mathbf{E}\left\{(\theta_n(i))^2 /\mathcal{F}_{n-1} \bigwedge u_n = u(i)\right\} \overset{a.s.}{=}$$

$$\overset{a.s.}{=} (1 - \lambda_n(i))^2 (\theta_{n-1}(i))^2 + (\lambda_n(i))^2 \sigma_n^2(i) \leq$$

$$\leq \left\{1 - \left[2 + O\left(\left[\sum_{t=1}^{n} \chi(u_t = u(i))\right]^{-1}\right)\right] \lambda_n(i)\right\} (\theta_{n-1}(i))^2 +$$

$$+ (\lambda_n(i))^2 \sigma^2$$

In view of Robbins-Siegmund theorem [23], it follows that $\theta_n(i) \underset{n\to\infty}{\overset{a.s.}{\to}} 0$ and hence

$$\lim_{n\to\infty} s_n(i) = \lim_{n\to\infty} \frac{\sum_{t=1}^{n} c_t(i)\chi(u_t = u(i))}{\sum_{t=1}^{n} \chi(u_t = u(i))} = \lim_{n\to\infty} \frac{1}{n}\sum_{t=1}^{n} c_t(i) = \tilde{c}(i)$$

So, (5.37) is proved.

3. (5.38) follows directly from (5.37) and assumption **(H3)**.

4. (5.39) is the consequence of (5.38). ■

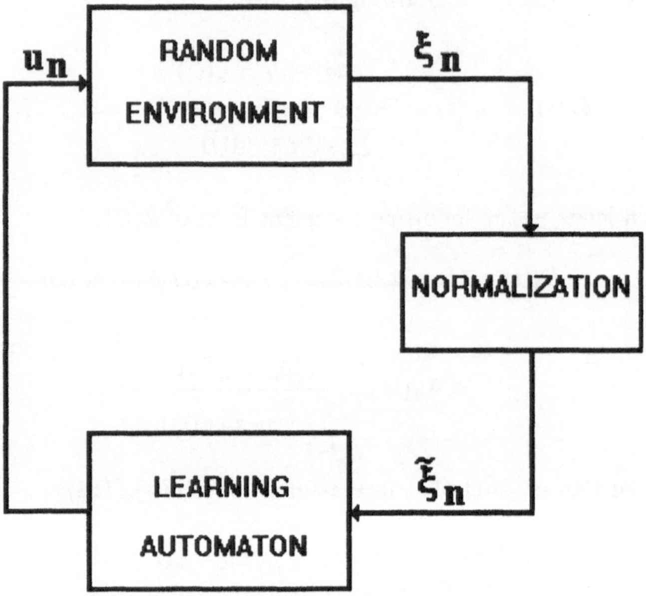

FIGURE 5.3. Learning automaton with normalization procedure.

In the following we shall be concerned with the behaviour of learning automata with normalized environment response. The probability distribution p_n will be adjusted using the previous reinforcement scheme (5.8) where ξ_n is replaced by $\widetilde{\xi}_n$ (see Figure 5.3).

To illustrate this approach, the analysis of the properties of Bush-Mosteller [18], Shapiro-Narendra [19] and Varshavskii-Vorontsova [20] reinforcement schemes is given in the following subsections. The correction factor was considered constant ($\gamma_n = \gamma = const.$) in the original version of these reinforcement schemes. In this study, we assume that the correction factor γ_n is selected according to (5.22) with $\gamma < (1 + a)$. The initial probabilities are assumed to be strictly positive
($p_1(i) > 0 \quad \forall i = 1, ..., N$).

5.7.1 Bush-Mosteller reinforcement scheme

The Bush-Mosteller scheme [18]-[5] is described by:

$$p_{n+1} = p_n + \gamma_n \left[e(u_n) - p_n + \widetilde{\xi}_n (e^N - N e(u_n))/(N-1) \right] \qquad (5.40)$$

where

$$\begin{aligned} \gamma_n &\in [0,1], \quad \widetilde{\xi}_n \in [0,1] \\ e(u_n) &= (\underbrace{0,...,0,1}_{i},0,...,0)^T, \; u_n = u(i) \\ e^N &= (1,...,1)^T \in R^N \end{aligned}$$

The reinforcement scheme above is frequently used in the case of stationary environments. We shall analyze its behaviour in asymptotically stationary (in the average sense) mediums. This question is addressed in the following theorem.

Theorem 1. *For the Bush-Mosteller scheme (5.40), condition (5.35) is satisfied, and if assumptions (H3) and (H4) hold, and the optimal action is single, i.e.,*

$$\min_{i \neq \alpha} \widetilde{\Delta}(i) := \widetilde{\Delta}^* > 0 \tag{5.41}$$

then, the properties (P1) and (P2) are fulfilled and the loss function Φ_n tends to its minimal possible value $\widetilde{c}(\alpha)$ (5.31), with probability one.

Proof.

Let us consider the following estimation:

$$p_{n+1}(i) = p_n(i) + \gamma_n[\chi(u_n = u(i)) - p_n(i) +$$

$$+\widetilde{\xi}_n\left[1 - N\chi(u_n = u(i))\right]/(N-1)] =$$

$$= p_n(i)\ (1-\gamma_n) +$$

$$+\frac{\gamma_n}{N-1}\left\{\widetilde{\xi}_n + \chi(u_n = u(i))\left[N(1-\widetilde{\xi}_n) - 1\right]\right\} \geq \tag{5.42}$$

$$\geq p_n(i)\ (1-\gamma_n) \geq \cdots \geq p_1(i)\prod_{t=1}^{n}(1-\gamma_t)$$

From lemma A.4 [5], it follows that

$$\prod_{t=1}^{n}(1-\gamma_t) \geq \begin{cases} \left(\frac{a-\gamma}{n+a}\right)^{\gamma}, & a > \gamma \in (0,1) \\ \frac{a}{n+a}, & \gamma = 1, \; a > 0 \end{cases} \tag{5.43}$$

Substitution of (5.43) into (5.42) leads to the desired result (5.35).

In view of (5.27) and assumptions **(H1)**, **(H2)** we deduce

$$\begin{aligned}\widetilde{\Delta}_n(\alpha) &= (1-\frac{1}{n})\,\widetilde{\Delta}_{n-1}(\alpha)+\frac{1}{n}\Delta_n \le \\ &\overset{a.s.}{\le} (1-\frac{1}{n})\,\widetilde{\Delta}_{n-1}(\alpha)+\frac{1}{n}const\end{aligned}$$

Using Lemma A.3-2 (see Appendix A) for

$$u_n=\widetilde{\Delta}_n(\alpha),\ \alpha_n=\frac{1}{n},\ \beta_n=\frac{1}{n}const,\ \nu_n=n^{1-\varepsilon}\ (\varepsilon\in(0,1))$$

we derive

$$\widetilde{\Delta}_n(\alpha)\overset{a.s.}{=}o(\frac{1}{n^{1-\varepsilon}}) \tag{5.44}$$

The reinforcement scheme (5.40) and (5.44) lead to

$$\begin{aligned}&\mathbf{E}\{[1-p_{n+1}(\alpha)]\,/\mathcal{F}_n\}\overset{a.s.}{=}[1-p_{n+1}(\alpha)]- \\ &-\gamma_n[\frac{1}{N-1}\sum_{i\neq\alpha}^{N}\mathbf{E}\left\{\widetilde{\xi}_n/\mathcal{F}_{n-1}\bigwedge u_n=u(i)\right\}\,p_n(i) \\ &-\mathbf{E}\left\{\widetilde{\xi}_n/\mathcal{F}_{n-1}\bigwedge u_n=u(i)\right\}\,p_n(i)]\le \\ &\le[1-p_{n+1}(\alpha)]-\gamma_n[\frac{1}{N-1}\sum_{i\neq\alpha}^{N}\left[\widetilde{\Delta}_n(i)+o_\omega(1)\right]\,p_n(i)- \\ &-\left[\widetilde{\Delta}_n(\alpha)+o_\omega(1)\right]\,p_n(\alpha)\le \\ &\le[1-p_{n+1}(\alpha)]\,(1-\gamma_n\frac{\widetilde{\Delta}^*}{N-1})+\gamma_n\,o(\frac{1}{n^{1-\varepsilon}})\end{aligned} \tag{5.45}$$

for $n\ge n_0(\omega)$, $n_0(\omega)\overset{a.s.}{<}\infty$.

Taking into account the normalization procedure ($\widetilde{\Delta}^*\in(0,1)$) (5.45) and Robbins-Siegmund theorem [23], we obtain

$$p_n(\alpha)\underset{n\to\infty}{\overset{a.s.}{\to}}1$$

The property **(P1)** is then fulfilled.

To prove the fulfillment of property **(P2)** we use Lemma 1 and Toeplitz lemma:

$$\lim_{n\to\infty}\frac{1}{n}\sum_{t=n_0}^{n}T_{t-1}\overset{a.s.}{=}\lim_{n\to\infty}\frac{1}{n}\sum_{t=n_0}^{n}\mathbf{E}\{T_{t-1}/\mathcal{F}_{t-1}\}\overset{a.s.}{=}$$

$$\stackrel{a.s.}{=} \lim_{n\to\infty} \mathbf{E}\{T_n/\mathcal{F}_n\} = \frac{e^N - Ne(u(\alpha))}{N-1}\sum_{i=1}^{N}\left[\widetilde{\Delta}_n(i)+o_\omega(1)\right]\ p_n(i) \stackrel{a.s.}{=}$$

$$\stackrel{a.s.}{=} \frac{e^N - Ne(u(\alpha))}{N-1}\widetilde{\Delta}(\alpha) = 0$$

Theorem is proved. ■

This theorem shows that, a learning automaton using the Bush-Mosteller reinforcement scheme with the normalization procedure described above, selects asymptotically the optimal action, i.e., its behaviour is asymptotically optimal.

The next subsection deals with the analysis of Shapiro-Narendra reinforcement scheme.

5.7.2 Shapiro-Narendra reinforcement scheme

The Shapiro-Narendra scheme [19]-[5] is described by:

$$p_{n+1} = p_n + \gamma_n(1-\widetilde{\xi}_n)\left[e(u_n) - p_n\right] \tag{5.46}$$

We shall analyze its behaviour in asymptotically stationary (in the average sense) mediums and state some analytical results. These results are described in the following theorem.

Theorem 2. *For the Shapiro-Narendra scheme [19] (5.46), assume that condition (5.35) is satisfied and suppose that assumptions (H1), (H2) and condition (5.41) hold. In addition, suppose that the correction factor satisfies the following condition*

$$\lim_{n\to\infty} \gamma_n \prod_{t=1}^{n}(1-\gamma_t)^{-1} := c < p_1(\alpha)\frac{\widetilde{\Delta}^*}{1-\widetilde{\Delta}^*} \tag{5.47}$$

then, the properties (P1) and (P2) are fulfilled and the loss function Φ_n tends to its minimal possible value $\widetilde{c}(\alpha)$, with probability one. (5.31)

Proof.

Let us estimate the lower bounds of the probabilities $p_{n+1}(i)$:

$$\begin{aligned} p_{n+1}(i) &= p_n(i) + \gamma_n(1-\widetilde{\xi}_n)\left[\chi(u_n = u(i)) - p_n(i)\right] = \\ &= p_n(i)\left[1-\gamma_n(1-\widetilde{\xi}_n)\right] + \gamma_n(1-\widetilde{\xi}_n)\chi(u_n = u(i)) \geq \\ &\geq p_n(i)\,(1-\gamma_n) \geq \cdots \geq p_1(i)\prod_{t=1}^{n}(1-\gamma_t) \end{aligned} \tag{5.48}$$

Substituting (5.43) into (5.48) leads to the desired result **(P1)**.

In view of (5.27) and assumptions **(H3)** and **(H4)**, and according to the proof of the previous theorem, we deduce that (5.44) is fulfilled.

Let us consider the following Lyapunov function

$$W_n := \frac{1 - p_n(\alpha)}{p_n(\alpha)}$$

Taking into account the Shapiro Narendra scheme and (5.44), it follows:

$$\mathbf{E}\{W_{n+1}/\mathcal{F}_n\} \overset{a.s.}{=} \sum_{i=1}^{N} \mathbf{E}\left\{W_{n+1}/\mathcal{F}_{n-1} \bigwedge u_n = u(i)\right\} p_n(i) =$$

$$= \sum_{i \neq \alpha}^{N} \left(\frac{1}{p_n(\alpha) - \gamma_n(1 - \left[\widetilde{\Delta}(i) + o_\omega(1)\right]) \, p_n(\alpha)} - 1 \right) p_n(i) +$$

$$+ \left(\frac{1}{p_n(\alpha) + \gamma_n(1 - \left[\widetilde{\Delta}(\alpha) + o_\omega(1)\right]) \, (1 - p_n(\alpha))} - 1 \right) p_n(\alpha) =$$

$$= \sum_{i \neq \alpha}^{N} \left(\frac{1}{p_n(\alpha)[1 - \gamma_n] + \gamma_n \widetilde{\Delta}(i) \, p_n(\alpha)} - 1 \right) p_n(i) +$$

$$+ \left(\frac{1}{p_n(\alpha) + \gamma_n \, (1 - p_n(\alpha))} - 1 \right) p_n(\alpha) + \gamma_n \, o_\omega(1)$$

Replacing $\widetilde{\Delta}(i)$ by its minimal value $\widetilde{\Delta}^*$ leads to the following inequality:

$$\mathbf{E}\{W_{n+1}/\mathcal{F}_n\} \overset{a.s.}{\leq} \left(\frac{1}{p_n(\alpha)[1 - \gamma_n] + \gamma_n \widetilde{\Delta}^* \, p_n(\alpha)} - 1 \right) (1 - \ p_n(\alpha)) +$$

$$+ \frac{(1 - \ p_n(\alpha)) \, p_n(\alpha)}{p_n(\alpha) + \gamma_n \, (1 - p_n(\alpha))} [1 - \gamma_n] + \gamma_n \, o_\omega(1) =$$

$$= W_n \left[\frac{1}{1 - \gamma_n + \gamma_n \widetilde{\Delta}^*} - p_n(\alpha) + \frac{p_n^2(\alpha)[1 - \gamma_n]}{p_n(\alpha)(1 - \gamma_n \ ) + \gamma_n} \right] + \gamma_n \, o_\omega(1) =$$

$$= W_n \left[\frac{1}{1 - \gamma_n + \gamma_n \widetilde{\Delta}^*} - \frac{\gamma_n}{(1 - \gamma_n \ ) + \gamma_n p_n^{-1}(\alpha)} \right] + \gamma_n \, o_\omega(1) \qquad (5.49)$$

for $n \geq n_0(\omega)$, $n_0(\omega) \overset{a.s.}{<} \infty$.

Condition (5.47) gives

$$\lim_{n \to \infty} \gamma_n p_n^{-1}(\alpha) \geq p_1^{-1}(\alpha) \lim_{n \to \infty} \gamma_n \prod_{t=1}^{n} (1 - \gamma_t)^{-1} = p_1^{-1}(\alpha) \, c \qquad (5.50)$$

Inequalities (5.50) and (5.49) imply

$$\mathbf{E}\left\{W_{n+1}/\mathcal{F}_n\right\} \overset{a.s.}{\leq} W_n \left[\frac{1}{1-\gamma_n+\gamma_n\widetilde{\Delta}^*} - \frac{\gamma_n}{(1-\gamma_n\ )+p_1^{-1}(\alpha)\ c+o(1)}\right]+$$

$$+\gamma_n\ o_\omega(1) =$$

$$= W_n\left(1+\gamma_n\left[1-\widetilde{\Delta}^* - \frac{1}{1+p_1^{-1}(\alpha)\ c}+o(1)\right]\right)+\gamma_n\ o_\omega(1) =$$

$$= W_n\left(1-\gamma_n\left[\frac{\widetilde{\Delta}^* - p_1^{-1}(\alpha)\ c\left(1-\widetilde{\Delta}^*\right)}{1+p_1^{-1}(\alpha)\ c}+o(1)\right]\right)+\gamma_n\ o_\omega(1)$$

Taking into account the normalization procedure ($\widetilde{\Delta}^* \in (0,1)$), assumption (5.47) and Robbins-Siegmund theorem [23], we obtain

$$W_n \underset{n\to\infty}{\overset{a.s.}{\to}} 0, \quad p_n(\alpha) \underset{n\to\infty}{\overset{a.s.}{\to}} 1$$

The property **(P1)** is then fulfilled.

To prove the fulfillment of property **(P2)** we use Lemma 1 and Toeplitz lemma:

$$\lim_{n\to\infty}\frac{1}{n}\sum_{t=n_0}^{n} T_{t-1} \overset{a.s.}{=} \lim_{n\to\infty}\frac{1}{n}\sum_{t=n_0}^{n}\mathbf{E}\left\{T_{t-1}/\mathcal{F}_{t-1}\right\} \overset{a.s.}{=}$$

$$\overset{a.s.}{=} \lim_{n\to\infty}\mathbf{E}\left\{T_n/\mathcal{F}_n\right\} = \sum_{i=1}^{N}\left[1-\widetilde{\Delta}(i)\right]\left[e(u(i)) - \lim_{n\to\infty} p_n\right]\lim_{n\to\infty} p_n(i) \overset{a.s.}{=}$$

$$\overset{a.s.}{=} e(u(\alpha)) - e(u(\alpha)) = 0$$

Theorem is proved. ■

The following corollary deals with the constraints associated with the correction factor γ_n.

Corollary 1. *For the correction factor (5.22), assumption (5.47) will be satisfied if*

$$\gamma = 1, \quad \lim_{n\to\infty}\gamma_n\prod_{t=1}^{n}(1-\gamma_t)^{-1} = c = a^{-1} < p_1(\alpha)\frac{\widetilde{\Delta}^*}{1-\widetilde{\Delta}^*}$$

These statements show that, a learning automaton using the Shapiro-Narendra reinforcement scheme with the normalization procedure described above, has an asymptotically optimal behaviour.

The Varshavskii-Vorontsova reinforcement scheme will be considered in the following.

5.7.3 VARSHAVSKII-VORONTSOVA REINFORCEMENT SCHEME

The Varshavskii-Vorontsova scheme [20]-[5] is described by:

$$p_{n+1} = p_n + \gamma_n p_n^T \, e(u_n)(1 - 2\widetilde{\xi}_n) \left[e(u_n) - p_n \right] \tag{5.51}$$

This scheme belongs to the class of nonlinear reinforcement schemes. We shall analyze its behaviour in asymptotically stationary (in the average sense) mediums. The following theorem cover all the specific properties of the Varshavskii-Vorontsova scheme. Its significance will be discussed below.

Theorem 3. *For the Varshavskii-Vorontsova scheme (5.51), condition (5.35) is satisfied and if assumptions (H3), (H4) and the condition (5.41) hold, and in addition if the nonstationary environment satisfies the following condition:*

$$b := \left(\sum_{i \neq \alpha}^{N} \left[1 - 2\widetilde{\Delta}(i) \right]^{-1} \right)^{-1} < 0 \tag{5.52}$$

then, the properties (P1) and (P2) are fulfilled and the loss function Φ_n tends to its minimal possible value $\widetilde{c}(\alpha)$, with probability one. (5.31)

Proof.

Let us estimate the lower bounds of the probabilities:

$$p_{n+1}(i) = p_n(i) + \gamma_n p_n^T \, e(u_n)(1 - 2\widetilde{\xi}_n)[\chi(u_n = u(i)) - p_n(i)] =$$

$$= p_n(i) \left[1 - \gamma_n p_n^T \, e(u_n)(1 - 2\widetilde{\xi}_n) \right] + \gamma_n p_n^T \, e(u_n) \chi(u_n = u(i)) \geq$$

$$\geq p_n(i) \left[1 - \gamma_n p_n^T \, e(u_n)(1 - 2\widetilde{\xi}_n) \right] \geq$$

$$\geq p_n(i) \left[1 - \gamma_n p_n^T \, e(u_n) \right] \geq$$

$$\geq p_n(i) \, (1 - \gamma_n) \geq \cdots \geq p_1(i) \prod_{t=1}^{n} (1 - \gamma_t) \tag{5.53}$$

Substituting (5.43) into (5.53) gives us the desired result (5.35).

For convergence analysis, let us introduce the following Lyapunov function:

$$W_n := \frac{1 - p_n(\alpha)}{p_n(\alpha)}$$

The reinforcement scheme (5.51), (5.44) and simple calculations lead to

$$\mathbf{E}\{W_{n+1}/\mathcal{F}_n\} \overset{a.s.}{=} \sum_{i=1}^{N} \mathbf{E}\left\{W_{n+1}/\mathcal{F}_{n-1} \bigwedge u_n = u(i)\right\} p_n(i) =$$

$$= \sum_{i \neq \alpha}^{N} \left(\frac{1}{p_n(\alpha) - \gamma_n (1 - 2\widetilde{\Delta}(i) + o_\omega(1))\, p_n(\alpha)\, p_n(i)} - 1 \right) p_n(i) +$$

$$+ \left(\frac{1}{p_n(\alpha) + \gamma_n (1 - 2\widetilde{\Delta}(\alpha) + o_\omega(1))\, (1 - p_n(\alpha))\, p_n(\alpha)} - 1 \right) p_n(\alpha) =$$

$$= \sum_{i \neq \alpha}^{N} \left(\frac{p_n(i)}{1 - \gamma_n \left[1 - 2\widetilde{\Delta}(i)\right] p_n(i)} \right) \frac{1}{p_n(\alpha)} + \tag{5.54}$$

$$+ \left(\frac{1}{1 + \gamma_n \, (1 - p_n(\alpha))} - 1 \right) + \gamma_n \, o_\omega(1)$$

for $n \geq n_0(\omega)$, $n_0(\omega) \overset{a.s.}{<} \infty$.

Let us now maximize the first term in (5.54) with respect to the components $p_n(i)$ $(i \neq \alpha)$ under the constraint

$$\sum_{i \neq \alpha}^{N} p_n(i) = 1 - p_n(\alpha) \tag{5.55}$$

To do that, let us introduce the following variables

$$x_i := p_n(i), \quad a_i := \gamma_n \left[1 - 2\widetilde{\Delta}(i)\right], \quad i \neq \alpha$$

It follows that

$$\sum_{i \neq \alpha}^{N} \left(\frac{p_n(i)}{1 - \gamma_n \left[1 - 2\widetilde{\Delta}(i)\right] p_n(i)} \right) = \sum_{i \neq \alpha}^{N} \frac{x_i}{1 - a_i x_i} := F(x) \tag{5.56}$$

To maximize the function $F(x)$ under the constraint (5.55), let us introduce the following Lagrange function:

$$L(x, \lambda) := F(x) - \lambda \left[\sum_{i \neq \alpha}^{N} x_i - (1 - x_\alpha) \right]$$

The optimal solution (x^*, λ^*) satisfies the following optimality conditions:

$$\frac{\partial}{\partial x_i} L(x^*, \lambda^*) = \frac{1}{(1 - a_i x_i^*)^2} - \lambda^* = 0 \quad \forall i \neq \alpha$$

$$\frac{\partial}{\partial \lambda} L(x^*, \lambda^*) = \sum_{i \neq \alpha}^{N} x_i^* - (1 - x_\alpha^*) = 0$$

From these optimality conditions, it can be seen that

$$x_i^* = \frac{1}{a_i} \frac{1 - x_\alpha}{\sum\limits_{i \neq \alpha}^{N} a_i^{-1}}, \quad \sqrt{\lambda^*} = (1 - \frac{1 - x_\alpha}{\sum\limits_{i \neq \alpha}^{N} a_i^{-1}})^{-1}$$

Hence

$$F(x) \leq F(x^*) = (1 - x_\alpha)(1 - \frac{1 - x_\alpha}{\sum\limits_{i \neq \alpha}^{N} a_i^{-1}})^{-1}$$

From this inequality we derive

$$\sum_{i \neq \alpha}^{N} \left(\frac{p_n(i)}{1 - \gamma_n \left[1 - 2\tilde{\Delta}(i)\right] p_n(i)} \right) \leq \frac{1 - p_n(\alpha)}{1 - \gamma_n b [1 - p_n(\alpha)]} \tag{5.57}$$

where

$$b := \left(\sum_{i \neq \alpha}^{N} \left[1 - 2\tilde{\Delta}(i)\right]^{-1} \right)^{-1} \tag{5.58}$$

Substituting (5.57) into (5.54), we obtain

$$\mathbf{E}\{W_{n+1}/\mathcal{F}_n\} \overset{a.s.}{\leq} W_n \frac{1}{1 - \gamma_n b [1 - p_n(\alpha)]} +$$

$$+ \left(\frac{1}{1 + \gamma_n (1 - p_n(\alpha))} - 1 \right) + \gamma_n \, o_\omega(1) =$$

$$= W_n \left[\frac{1}{1 - \gamma_n b [1 - p_n(\alpha)]} + \frac{\gamma_n \, p_n(\alpha)}{1 + \gamma_n (1 - p_n(\alpha))} \right] + \gamma_n \, o_\omega(1) =$$

$$= W_n \left[1 - \gamma_n (|b| [1 - p_n(\alpha)] + p_n(\alpha)) + O(\gamma_n^2)\right] + \gamma_n \, o_\omega(1)$$

The maximization of the right side of this inequality with respect to $p_n(\alpha) \in [0, 1]$ leads to

$$\mathbf{E}\{W_{n+1}/\mathcal{F}_n\} \overset{a.s.}{\leq} W_n \left[1 - \gamma_n \min\{|b|; 1\} + O(\gamma_n^2)\right] + \gamma_n \, o_\omega(1)$$

Taking into account condition (5.52) and in view of Robbins-Siegmund theorem [23], we obtain

$$W_n \underset{n\to\infty}{\overset{a.s.}{\to}} 0, \quad p_n(\alpha) \underset{n\to\infty}{\overset{a.s.}{\to}} 1$$

The property **(P1)** is then fulfilled.

Using Toeplitz lemma gives us

$$\lim_{n\to\infty} \frac{1}{n} \sum_{t=n_0}^{n} T_{t-1} \overset{a.s.}{=} \lim_{n\to\infty} \frac{1}{n} \sum_{t=n_0}^{n} \mathbf{E}\{T_{t-1}/\mathcal{F}_{t-1}\} \overset{a.s.}{=} \lim_{n\to\infty} \mathbf{E}\{T_n/\mathcal{F}_n\}$$

Then, by using Lemma 1, it follows:

$$\lim_{n\to\infty} \mathbf{E}\{T_n/\mathcal{F}_n\} = \sum_{i=1}^{N} \left[1 - 2\widetilde{\Delta}(i)\right] \lim_{n\to\infty} p_n^2(i) \left[e(u(i)) - \lim_{n\to\infty} p_n\right] \overset{a.s.}{=}$$

$$\overset{a.s.}{=} \left[1 - 2\widetilde{\Delta}(\alpha)\right] \left[e(u(\alpha)) - e(u(\alpha))\right] = 0$$

Property (P2) is fulfilled. Theorem is proved. ■

The following corollary gives an example of the class of environments satisfying condition (5.52).

Corollary 2. *For example, condition (5.52) will be satisfied for the class of environments for which*

$$\widetilde{\Delta}(i) > \frac{1}{2} \quad \forall i \neq \alpha$$

The main point of this theorem is that, a learning automaton using the Varshavskii-Vorontsova reinforcement scheme with the normalization procedure described above, has an asymptotically optimal behaviour in the class of nonstationary environments for which the condition (5.52) is fulfilled.

The next section presents some simulation results.

5.8 Simulation results

This section presents a simple numerical simulation, from which we can verify the viability of the design and analysis given in this chapter. Let us consider the optimization of a function described by (5.5) with $c = 10$, $u^* = 0.01$ and $\omega = 10$. The variation domain of the variable u is discretised into a set of N values

$$[u(1),\ u(2), \ldots,\ u(10)]\,;\ u(i) = 5 + i$$

No prior information is used initially, i.e., the probability vector is initialized to $p_0 = \left[\frac{1}{N}, ..., \frac{1}{N}\right]^T$.

The learning system operates as follows: At each time n a normally distributed random variable z is generated and an action $u(i)$ is selected on the basis of the probability distribution p_n. The selected action $u(i)$ is then used to construct the normalized automaton input which in turn is used to adjust the probability distribution by means of a reinforcement scheme (Bush-Mosteller (B-M), Shapiro-Narendra (S-N) and Varshavskii-Vorontsova (V-V)). This procedure is repeated at each time.

The optimal action is $u(5)$ and is equal to 10. The probabilities associated with this optimal action (solution) and generated respectively by Bush-Mosteller (B-M), Shapiro-Narendra (S-N) and Varshavskii-Vorontsova (V-V) reinforcement schemes are given in Figure 5.4.

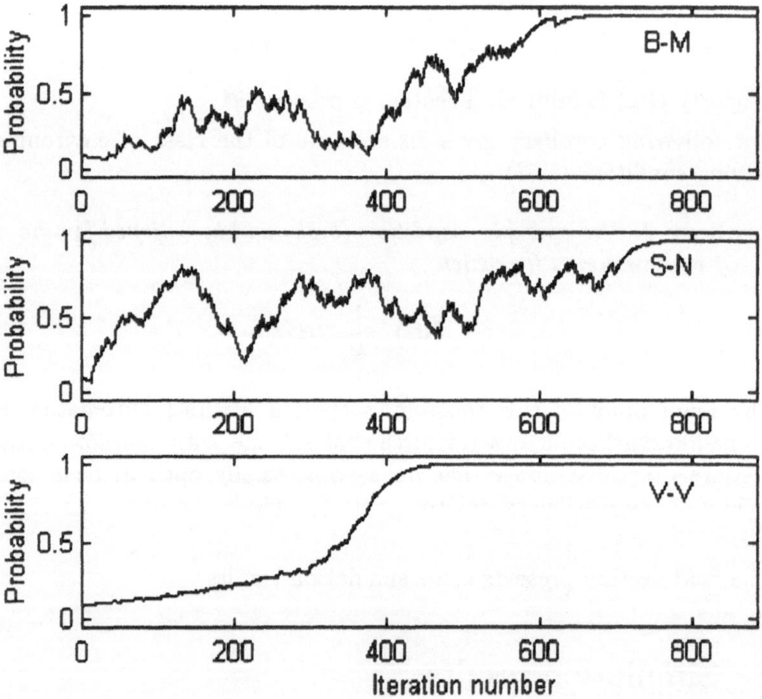

FIGURE 5.4. Evolution of the probabilities associated with the optimal action.

After a learning period which is equal to:
450 iterations for the Varshavskii-Vorontsova scheme
650 iterations for the Bush-Mosteller scheme
750 iterations for the Shapiro-Narendra scheme,

these probabilities tend to one. The loss functions associated with the behaviour of learning automata using different reinforcement schemes are depicted in Figure 5.5.

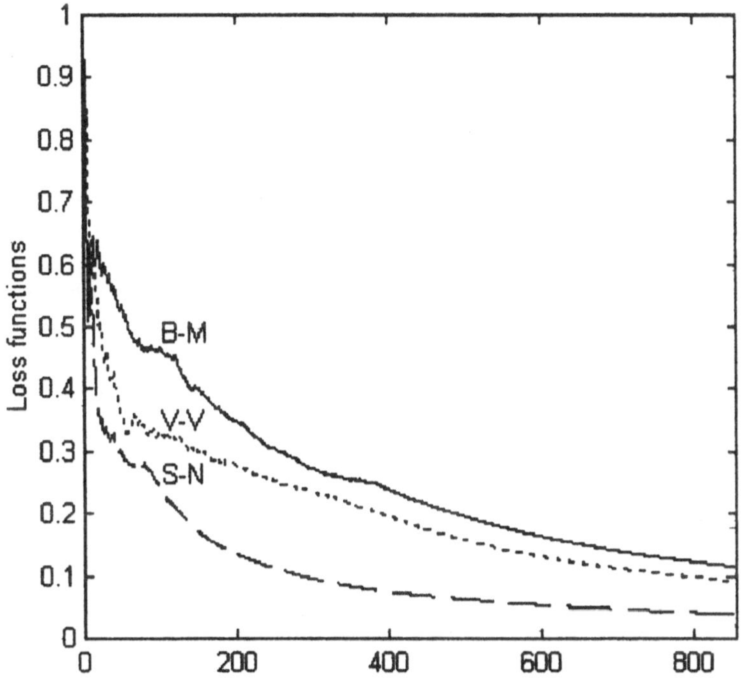

FIGURE 5.5. Evolution of the loss functions.

The three reinforcement schemes considered in this study achieve a satisfactory decrease in the loss function value. The Shapiro-Narendra scheme (S-N) leads to a loss function which decreases more quickly than those related respectively to Bush-Mosteller (B-M) and Varshavskii-Vorontsova (V-V) reinforcement schemes. Figures 5.4 and 5.5 illustrate the ability the optimization algorithm to deal with the optimization of nonstationary functions. It can be seen that the optimization objective is achieved. This behaviour was expected from the theoretical results stated in this chapter. The above results can be explained by the learning ability of the optimization algorithm.

5.9 Conclusion

This chapter has focused on the use of variable-structure stochastic learning automata operating in nonstationary environments for solving optimization problems related to nonstationary functions. The conditional expectation of the environment responses (automaton inputs) were assumed to be time-varying. Different nonstationary environments (Markovian environments, periodically changing environments, multi-Teacher environments, environments with fixed optimal actions) have been analyzed. A normalization procedure has been introduced to deal with environment responses which do not belong to the unit segment. The notion of asymptotically stationary environment (in the average sense) was also introduced. Several theoretical results are stated. These results concern the asymptotically optimal behaviour of learning automata using Bush-Mosteller, Shapiro-Narendra and Varshavskii-Vorontsova reinforcement schemes. Some simulation results have been presented to illustrate the use of learning automata operating in nonstationary environments for solving optimization problems related to nonstationary functions. The theoretical analysis presented in this chapter can be extended to learning automata which use other reinforcement schemes.

References

[1] Holst J 1977 *Adaptive Prediction and Recursive Estimation*, Department of Automatic Control, Lund Institute of Technology. Report LUTFD2/(TFRT-1013)/1-206/

[2] Najim K 1989 *Process Modeling and Control in Chemical Engineering.* Marcel Dekker, New York

[3] Kawashima H 1981 Identification of autoregressive integrated processes. IFAC Congress, Japan, Kyoto

[4] Yaglom A M 1962 *An Introduction to the Theory of Stationary Random Functions.* Prentice-Hall, Englewood Cliffs

[5] Najim K, Poznyak A S 1994 *Learning Automata: Theory and Applications.* Pergamon Press, Oxford

[6] Tseltin M L 1973 *Automaton Theory and Modeling of Biological Systems.* Academic Press, New York

[7] Narendra K S, Thathachar M A L 1989 *Learning Automata an Introduction.* Prentice-Hall, Englewood Cliffs

[8] Najim K, Oppenheim G 1991 Learning systems: theory and applications. *IEE Proceedings Computer and Digital Techniques* 138:183-192

[9] Srikantakumar P R, Narendra K S 1982 A learning model for routing in telephone Networks. *SIAM J. Control and Optimization* 20:34-57

[10] Barto A, Anandan P, Anderson C W 1986 Cooperatively in networks of pattern recognizing stochastic learning automata. In Narendra K S (ed) 1986 *Adaptive and Learning Systems : Theory and Applications.* Plenum Press, New York

[11] Baba N, Mogami Y 1988 Learning behaviours of hierarchical structure of stochastic automata in a nonstationary multi-teacher environment. *Int. J. Systems Sci.* 19:1345-1350

[12] Baba N 1984 *New Topics in Learning Automata Theory and Applications.* Springer-Verlag, Berlin

[13] Narendra K S, Viswanathan R 1972 A two-level system of stochastic automata for periodic random environments. *IEEE Trans. Syst. Man, and Cybern.* 2:285-289

[14] Baba N, Sawaragi Y 1975 On the learning behaviour of stochastic automata under a nonstationary random environment. *IEEE Trans. Syst. Man, and Cybern.* 5:273-275

[15] Narendra K S, Thathachar M A L 1980 On the behavior of learning automata in a changing environment with application to telephone traffic routine. *IEEE Trans. Syst. Man, and Cybern.* 10:262-269

[16] Nedzelnitsky O V Jr, Narendra K S 1987 Nonstationary models of learning automata routing in data communication networks. *IEEE Trans. Syst. Man, and Cybern.* 17:1004-1015

[17] Thathachar M A L, Ramakrishnan K R 1981 A hierarchical system of learning automata. *IEEE Trans. Syst. Man, and Cybern.* 11:236-241

[18] Bush R R, Mosteller F 1958 *Stochastic Models for Learning.* John Wiley & Sons, New York

[19] Shapiro I J, Narendra K S 1969 Use of stochastic automata for parameter self optimization with multimodal performance criteria. *IEEE Trans. Syst. Man, and Cybern.* 5:352-361

[20] Varshavskii V I, Vorontsova I P 1963 On the behavior of stochastic automata with variable structure. *Automation and Remote Control* 24:327-333

[21] Chandrasekaran B,and Shen D W C 1967 Adaptation of stochastic automata in non-stationary environment. *Proc. Natl. Electronics Conf.* pp. 39-44

[22] Doob J L 1953 *Stochastic Processes.* John Wiley & Sons, New York

[23] Robbins H, Siegmund D 1971 A convergence theorem for nonnegative almost supermartingales and some applications. In Rustagi J S (ed) 1971 *Optimizing Methods in Statistics.* Academic Press, New York

Appendix A: **Theorems and Lemmas**

This appendix collects the theorems and lemmas needed for the proof of several theorems which have been established in the previous chapter. Each lemma will be denoted by **A.i-j**, where **i** represents the chapter's number and **j** the order of classification in the considered chapter. The references cited in these appendices are given the bibliography section of chapter 1.

Theorem *(Kall's theorem, 1966). For the linear equation*

(K1)

$$Ax = b \ \left(A \in R^{m \times n}, b \in R^m, n \geq m+1\right)$$

where the matrix A has the first m independent columns, written $A_1, ..., A_m$, to have a nonnegative solution $x \in R^n$ for all $b \in R^m$, it is necessary and sufficient that there exist

(K2)

$$\mu_j \geq 0, \ \lambda_j < 0$$

such that

(K3)

$$\sum_{j=m+1}^{n} \mu_j A_j = \sum_{i=1}^{m} \lambda_j A_j$$

where A_j is the j^{-th} column of the matrix A.

Remark 5 *In fact, the matrix A must have more than m columns, because it is impossible that some $b = b_0$ and also $-b_0$ (such situation is possible) be represented as a nonnegative linear combination the same m independent columns. Hence, $n \geq m+1$.*

Proof.

a) *Necessity.*

The vectors $A_1, ..., A_m$ are independent, then there exit a numbers βi $(i = 1, ..., m)$ such that

(K4)

$$b = \sum_{i=1}^{m} \beta_j A_j$$

Then, for any $x \in R^n$ $(x_i \geq 0)$ satisfying (K1) we can write:

$$Ax = \sum_{j=1}^{n} x_j A_j = \sum_{i=1}^{m} \beta_j A_j$$

or

$$\sum_{i=1}^{m}(\beta_i - x_i)\, A_i = \sum_{j=m+1}^{n} x_j A_j$$

It is clear that if we want to prove this statement for any $b \in R^m$ there exist such b that for the given matrix A the numbers β_j will be nonpositive and hence we can identify μ_j in (K3) with the coefficients $(\beta_i - x_i)$ of the left side of the last inequality and λ_j in (K3) with the coefficients in the right side, i.e.,

$$\begin{aligned} \mu_j &= \beta_i - x_i \\ \lambda_j &= x_j \end{aligned}$$

So, the necessity is proved.

2) *Sufficiency.*

Taking into account the presentation (K4) and the linear independence of the vectors $A_1, ..., A_m$ we can conclude the for any fixed b the corresponding values β_j are unique. If all of them are nonnegative then the equation (K1) has the solution

$$x_j = \begin{cases} \beta_j & \text{for } j = 1, ..., m \\ 0 & \text{for } j = m+1, ...n \end{cases}$$

Let us now suggest that one at least of the β_j is negative. Notice that the largest of the ratios β_j/λ_j $(j \leq m)$ is positive, if at least one of the β_j founded above is negative. Let us define γ_0 as follows

$$\gamma_0 := \max_j |\beta_j| / |\lambda_j|$$

If relation (K3) is true then the following relation

$$\sum_{j=m+1}^{n} \gamma_0 \mu_j A_j = \sum_{i=1}^{m} \gamma_0 \lambda_j A_j$$

is also true, and hence, taking into account the representation (K4) we derive:

$$\sum_{i=1}^{m}(\gamma_0\lambda_j - \beta_j + \beta_j)\, A_j = \sum_{i=1}^{m}(\gamma_0\lambda_j - \beta_j)\, A_j + b =$$

$$= \sum_{j=m+1}^{n} \gamma_0 \mu_j A_j$$

or

$$b = \sum_{i=1}^{m}(\gamma_0 |\lambda_j| + \beta_j)\, A_j + \sum_{j=m+1}^{n} \gamma_0 \mu_j A_j =$$

$$= \sum_{i=1}^{m} |\lambda_j| \left(\gamma_0 + \frac{\beta_j}{|\lambda_j|}\right) A_j + \sum_{j=m+1}^{n} \gamma_0 \mu_j A_j$$

Taking into account that

$$\gamma_j := \gamma_0 + \frac{\beta_j}{|\lambda_j|} \geq 0$$

Finally, we can write

$$b = \sum_{i=1}^{m} |\lambda_j| \gamma_j A_j + \sum_{j=m+1}^{n} \gamma_0 \mu_j A_j =$$

$$= \sum_{j=1}^{n} x_j A_j$$

with

$$x_j = \begin{cases} |\lambda_j| \gamma_j & \text{for } j = 1, ..., m \\ \gamma_0 \mu_j & \text{for } j = m+1, ..., n \end{cases} \geq 0$$

Theorem is proved. ■

Lemma A.3-1. *Let $\{\mathcal{F}_n\}$ be a sequence of σ-algebras and η_n, θ_n, λ_n, and ν_n are $\mathcal{F}_n$-measurable nonnegative random variables such that*

1. $\mathbf{E}(\eta_n) < \infty$
2. $\sum_{n=1}^{\infty} \mathbf{E}(\theta_n) < \infty$
3. $\sum_{n=1}^{\infty} \lambda_n \overset{a.s.}{=} \infty, \quad \sum_{n=1}^{\infty} \nu_n \overset{a.s.}{<} \infty$
4. $\mathbf{E}(\eta_{n+1}/\mathcal{F}_n) \overset{a.s.}{\leq} (1 - \lambda_{n+1} + \nu_{n+1})\eta_n + \theta_n$

Then,

$$\lim_{n \to \infty} \eta_n \overset{a.s.}{=} 0$$

Proof.

In view of the assumptions of the previous theorem and Robbins-Siegmund theorem (Robbins and Siegmund, 1971), it follows that

$$\eta_n \underset{n \to \infty}{\overset{a.s.}{\to}} \eta^*$$

and

$$\sum_{n=1}^{\infty} \lambda_{n+1} \eta_n \overset{a.s.}{<} \infty$$

As

$$\sum_{n=1}^{\infty} \lambda_n \overset{a.s.}{=} \infty$$

Then, a subsequence η_{n_k} which tends to zero with probability 1 exists.

Hence $\eta^* \overset{a.s.}{=} 0$. ■

Lemma A.3-2. *Let $\{u_n\}$ be a sequence of nonnegative random variables u_n measurable with respect to the σ-algebra $\mathcal{F}_n$, for all $n = 1, 2...,$. If*

1. $\mathbf{E}(u_n/\mathcal{F}_n)$ $\forall$ $n = 1, 2, ...$ *exists*
2. *the following inequality holds*

$$\mathbf{E}(u_{n+1}/\mathcal{F}_n) \leq u_n(1 - \alpha_n) + \beta_n$$

where $\{\alpha_n\}$ and $\{\beta_n\}$ are sequences of non random variables such that

$$\alpha_n \in (0, 1], \quad \beta_n \geq 0$$

$$\sum_{n=1}^{\infty} \alpha_n = \infty, \sum_{n=1}^{\infty} \beta_n \nu_n < \infty$$

for some nonnegative sequence $\{\nu_n\}$ $(\nu_n > 0, n = 1, 2, ...)$

3. *the limit*

$$\lim_{n \to \infty} \frac{\nu_{n+1} - \nu_n}{\alpha_n \nu_n} := \mu < 1$$

exists,

then,

$$u_n = o_\omega(\frac{1}{\nu_n}) \overset{a.s.}{\to} 0$$

with probability 1 when $\nu_n \to \infty$.

Proof.

Let $\widetilde{u}_n$ be the sequence defined as

$$\widetilde{u}_n := u_n \nu_n$$

Then, using assumption (2), we obtain

$$\mathbf{E}(\widetilde{u}_{n+1}/\mathcal{F}_n) \overset{a.s.}{\leq} \widetilde{u}_n(1 - \alpha_n)(\frac{\nu_{n+1}}{\nu_n}) + \nu_{n+1}\beta_n =$$

$$= \widetilde{u}_n(1 - \alpha_n)(\frac{\nu_{n+1} - \nu_n}{\nu_n} - 1) + \nu_{n+1}\beta_n$$

Taking into account assumption (3), we derive

$$\mathbf{E}(\widetilde{u}_{n+1}/\mathcal{F}_n) \overset{a.s.}{\leq} \widetilde{u}_n \left[1 - \alpha_n(1 - \mu + o(1))\right] + \nu_{n+1}\beta_n$$

From this inequality and Robbins-Siegmund theorem (Robbins and Siegmund, 1971), we obtain

$$\widetilde{u}_n \overset{a.s.}{\to} 0$$

which is equivalent to

$$u_n = o_\omega\left(\frac{1}{\nu_n}\right) \overset{a.s.}{\to} 0$$

Lemma is proved. ■

Lemma A.3-3 (see ref. 31 of chapter 2). *For the sequence* $\{\gamma_n\}$ *where*

$$\gamma_n = \frac{\gamma}{n+a},$$

the following inequalities are fulfilled

a) for $\gamma \in (0,1)$ *and* $a > \gamma$, *then*

$$\left(\frac{1+a}{n+a-\gamma+1}\right)^\gamma \geq \prod_{k=1}^{n}(1-\gamma_k) \geq \left(\frac{a-\gamma}{n+a}\right)^\gamma$$

b) for $\gamma = 1$ *and* $a > 0$, *then*

$$\prod_{k=1}^{n}(1-\gamma_k) = \frac{a}{n+a}$$

Proof.

The proof of assertion a) is evident. Using the convexity property of the function $x \ln x$, it follows that

$$(x + \Delta x)\ln(x + \Delta x) - x\ln x \geq \Delta x(1 + x\ln x)$$

Taking into account this inequality, we obtain

$$\left(\frac{1+a}{n+a-\gamma+1}\right)^\gamma = \exp\left\{\int_1^{n+1} \ln\left(1 - \frac{\gamma}{n+a}\right) dx\right\} \geq$$

$$\geq \prod_{k=1}^{n}(1-\gamma(k)) = \exp\left\{\sum_{k=1}^{n} \ln\left(1 - \frac{\gamma}{k+a}\right) dx\right\} \geq$$

$$\geq \exp\left\{\int_{0}^{n+1} \ln\left(1-\frac{\gamma}{x+a}\right) dx\right\} = \left(\frac{a-\gamma}{n+a}\right)^{\gamma}$$

Lemma is proved. ■

Lemma A.4-1. *Under assumptions (H1) and (H2)*

1) *the random variables $\Phi^j(\omega)$ $(j = 0,...,m)$ are the partial limits of the sequences $\{\Phi_n^j\}$ $(j = 0,...,m)$ for almost all $\omega \in \Omega$ if and only if they can be written in the following form*

$$\Phi^0(\omega) = \sum_{i=1}^{N} v_i^0 p(i) := V_0(p)$$

$$\Phi^j(\omega) = \sum_{i=1}^{N} v_i^j p(i) := V_j(p), \quad (j = 1,...,m)$$

where the random vector $p = p(\omega) \in S^N$ (defined by equation (4.11)) is a limit point of the vector sequence $f_n = (f_n(1),...,f_n(N))^T$ defined by (equation (4.13)).

2) *for almost all $\omega \in \Omega$*

$$\Phi^j(\omega) \in \left[\min_i v_i^j, \max_i v_i^j\right] \quad (j = 0,...,m)$$

Proof.

Let us rewrite Φ_n^j (equation (4.1)) in the following form

$$\Phi_n^j = \sum_{i=1}^{N} f_n(i)\overline{\zeta}_n^j(u_n,\omega)$$

where

$$\overline{\zeta}_n^j(u_n,\omega) = \begin{cases} \dfrac{\sum_{t=1}^{n} \chi(u_t = u(i))\zeta_n^j(u(i),\omega)}{\sum_{t=1}^{n} \chi(u_t = u(i))} & , \quad \sum_{t=1}^{n} \chi(u_t = u(i)) > 0 \\ 0 & , \quad \sum_{t=1}^{n} \chi(u_t = u(i)) = 0 \end{cases}$$

are the current average loss functions for $u_t = u(i)$ $(i = 1,...,N)$. According to lemma A.12 (Najim and Poznyak, 1994) (see also Ash, 1972), for almost all

$$\omega \in \mathcal{B}_i := \left\{\omega \mid \sum_{t=1}^{n} \chi(u_t = u(i)) = \infty\right\}$$

we have

$$\lim_{n\to\infty} \overline{\zeta}_n^j(u_n,\omega) = v^i$$

For almost all $\omega \notin \mathcal{B}_i$, we have:

$$\lim_n \left|\overline{\zeta}_n^j(u_n,\omega)\right| < \infty$$

and

$$\lim_{n\to\infty} f_n(i) = 0$$

The vector f_n also belongs to the simplex S^N. It follows that any partial limit $\Phi^j(\omega)$ of a sequence $\{\Phi_n^j\}$ can be expressed in the following form:

$$\Phi^j(\omega) = \sum_{i=1}^{N} v_i^j p(i) := V_j(p), \quad (j = 0, ..., m)$$

where p is a partial limit of the sequence $\{f_n\}$ and consequently,

$$\Phi^j(\omega) \in \left[\min_i v_i^j, \max_i v_i^j\right] \ (j = 0, ..., m)$$

Lemma is proved. ■

Lemma A.4-2. *Let us assume that*

1. *the control strategy $\{u_n\}$ is stationary, i.e.,*

$$\mathbf{P}\ \{\omega : u_n = u(i) | \mathcal{F}_{n-1}\} \overset{a.s.}{=} p(i)$$

2. *the random variables $\zeta_n^j(u(i),\omega)$ $(j = 0, ..., m; i = 1, ..., N)$ have stationary conditional mathematical expectation and uniformly bounded conditional second moment, i.e.,*

$$\mathbf{E}\left\{\zeta_n^j(u(i),\omega) \mid u_n = u(i) \wedge \mathcal{F}_{n-1}\right\} \overset{a.s.}{=} v_i^j$$

$$\sup_n \mathbf{E}\left\{\left(\zeta_n^j(u(i),\omega)\right)^2 \mid u_n = u(i) \wedge \mathcal{F}_{n-1}\right\} \overset{a.s.}{<} \infty$$

3. $$\sum_{t=1}^{\infty} \chi(u_t = u(i)) \overset{a.s.}{\to} \infty$$

Then,

$$s_n^{ij} := \frac{\sum_{t=1}^{n} \zeta_t^j(u(i),\omega)\chi(u_t = u(i))}{\sum_{t=1}^{n} \chi(u_t = u(i))} - v_i^j \overset{a.s.}{\to} 0$$

Proof.

From the recurrent form of s_n^{ij}

$$s_n^{ij} = s_{n-1}^{ij}\left(1 - \frac{\chi(u_n = u(i))}{\sum_{t=1}^{n}\chi(u_t = u(i))}\right) +$$

$$+\frac{\chi(u_n = u(i))}{\sum_{t=1}^{n}\chi(u_t = u(i))}\left(\zeta_n^j(u(i),\omega) - v_i^j\right)$$

we derive

$$\mathbf{E}\left\{\left(s_n^{ij}\right)^2 \mid u_n = u(i) \wedge \mathcal{F}_{n-1}\right\} \overset{a.s.}{\leq} \left(s_{n-1}^{ij}\right)^2 \times$$

$$\times\left(1 - \frac{2\chi(u_n = u(i)) + o_\omega(1)}{\sum_{t=1}^{n}\chi(u_t = u(i))}\right) + \frac{\chi(u_n = u(i))}{\left[\sum_{t=1}^{n}\chi(u_t = u(i))\right]^2}\times$$

$$\times\mathbf{E}\left\{\left(\zeta_n^j(u(i),\omega)\chi(u_n = u(i)) - v_i^j\right)^2 \mid u_n = u(i) \wedge \mathcal{F}_{n-1}\right\} \leq$$

$$\leq \left(s_{n-1}^{ij}\right)^2\left(1 - \frac{\chi(u_n = u(i))\,(2 + o_\omega(1))}{\sum_{t=1}^{n}\chi(u_t = u(i))}\right) +$$

$$+\frac{\chi(u_n = u(i))}{\left[\sum_{t=1}^{n}\chi(u_t = u(i))\right]^2}Const(\omega)$$

where $o_\omega(1)$ is a random sequence tending to zero with probability one, and $Const(\omega)$ is an almost surely bounded and positive random variable .

Observe that

$$\sum_{n=n_0}^{\infty}\frac{\chi(u_n = u(i))\,(2 + o_\omega(1))}{\sum_{t=1}^{n}\chi(u_t = u(i))} \overset{a.s.}{=} \infty$$

In view of Robbins-Siegmund theorem (Robbins and Siegmund, 1971) (or Lemma A.9 in Najim and Poznyak, 1994), the previous inequality leads to the desired result. Lemma is proved. ■

The lemma used in chapter 5 is stated and proved in the following.

Lemma A.5-1 *(the matrix version of the Abel's identity).*

$$\sum_{t=n_0}^{n} A_t B_t = A_n \sum_{t=n_0}^{n} B_t - \sum_{t=n_0}^{n} [A_t - A_{t-1}] \sum_{s=n_0}^{t-1} B_s$$

$$A_t \in R^{m \times k}, \quad B_t \in R^{k \times l}$$

Proof.

For $n = n_0$ we obtain:

$$A_{n_0} B_{n_0} = A_{n_0} B_{n_0} - [A_{n_0} - A_{n_0-1}] \sum_{s=n_0}^{n_0-1} B_s = A_{n_0} B_{n_0}$$

The sum $\sum_{s=n_0}^{n_0-1} B_s$ in the previous equality is zero by virtue of the fact that the upper limit of this sum is less than the lower limit.

We use induction. We note that the identity (Abel's identity) is true for n_0. We assume that it is true for n and prove that it is true for $n+1$:

$$\sum_{t=n_0}^{n+1} A_t B_t = \sum_{t=n_0}^{n} A_t B_t + A_{n+1} B_{n+1} =$$

$$= A_n \sum_{t=n_0}^{n} B_t - \sum_{t=n_0}^{n} [A_t - A_{t-1}] \sum_{s=n_0}^{t-1} B_s + A_{n+1} B_{n+1} =$$

$$= \left(A_{n+1} \sum_{t=n_0}^{n} B_t + A_{n+1} B_{n+1} \right) -$$

$$- \left((A_{n+1} - A_n) \sum_{t=n_0}^{n} B_t + \sum_{t=n_0}^{n} [A_t - A_{t-1}] \sum_{s=n_0}^{t-1} B_s \right) =$$

$$= A_{n+1} \sum_{t=n_0}^{n+1} B_t - \sum_{t=n_0}^{n+1} [A_t - A_{t-1}] \sum_{s=n_0}^{t-1} B_s$$

The identity (Abel's identy) is proved. ■

Appendix B: **Stochastic Processes**

In this appendix we shall review the important definitions and some properties concerning stochastic processes.

A stochastic process, $\{x_n, n \in N\}$ is a collection (family) of random variables indexed by a real parameter n and defined on a probability space $(\Omega, \mathcal{F}, \mathbf{P})$ where Ω is the space of elementary events ω, $\mathcal{F}$ the basic σ-algebra and $\mathbf{P}$ the probability measure. A σ-algebra $\mathcal{F}$ is a set of subsets of Ω (collection of subsets). $\mathcal{F}(x_n)$ denotes the σ-algebra generated by the set of random variables x_n. The σ-algebra represents the knowledge about the process at time n. A family $\mathcal{F} = \{\mathcal{F}_n, n \geq 0\}$ of σ-algebras satisfy the standard conditions $\mathcal{F}_s \leq \mathcal{F}_n \leq \mathcal{F}$ for $s \leq n$, $\mathcal{F}_0$ is suggested by sets of measure zero of $\mathcal{F}$, and $\mathcal{F}_n = \bigcap_{s \geq n} \mathcal{F}_s$.

Let $\{x_n\}$ be a sequence of random variables with distribution function $\{F_n\}$ we say that:

Definition 7 *$\{x_n\}$ converges in distribution (law) to a random variable with distribution function F if the sequence $\{F_n\}$ converges to F.*

This is written

$$x_n \xrightarrow{law} x$$

Definition 8 *$\{x_n\}$ converges in probability to a random variable x if given ε, $\delta > 0$, then there exists $n_0(\varepsilon, \delta)$ such that $\forall n > n_0$ $P(\mid x_n - x \mid > \varepsilon) < \delta$*

This is written

$$x_n \xrightarrow{prob} x$$

Definition 9 *$\{x_n\}$ converges almost surely (with probability 1) to a random variable x if given ε, $\delta > 0$, then there exists $n_0(\varepsilon, \delta)$ such that $\forall n > n_0$. $P(\mid x_n - x \mid < \varepsilon \ \forall n > n_0) > 1 - \delta$*

This is written

$$x_n \xrightarrow{a.s.} x$$

Definition 10 $\{x_n\}$ *converges in quadratic mean to a random variable* x *if*

$$\lim_{n \to \infty} \mathbf{E}\left[(x_n - x)^T (x_n - x)\right] = 0$$

This is written

$$x_n \overset{q.m.}{\to} x$$

The relationships between these convergence concepts are summarized in the following

1. convergence in probability implies convergence in law.
2. convergence in quadratic mean implies convergence in probability.
3. convergence almost surely implies convergence in probability.

In general, the converse of these statements is false.

Stochastic processes as martingales have extensive applications in stochastic problems. They arise naturally whenever one needs to consider mathematical expectations with respect to increasing information patterns. They will be used to state several theoretical results concerning the convergence and the convergence rate of learning systems.

Definition 11 *A sequence of random variables* $\{x_n\}$ *is said to be adapted to a the sequence of increasing* σ*-algebras* $\{\mathcal{F}_n\}$ *if* x_n *is* $\mathcal{F}_n$ *measurable for every* n.

Definition 12 *A stochastic process* $\{x_n\}$ *is a martingale if*

$$E\{| x_n |\} \overset{a.s.}{<} \infty$$

and

$$E\{x_{n+1}/\mathcal{F}_n\} \overset{a.s.}{=} x_n$$

Definition 13 *A stochastic process* $\{x_n\}$ *is a supermartingale if*

$$E\{x_{n+1}/\mathcal{F}_n\} \overset{a.s.}{\leq} x_n$$

Definition 14 *A stochastic process* $\{x_n\}$ *is a submartingale if*

$$E\{x_{n+1}/\mathcal{F}_n\} \overset{a.s.}{\geq} x_n$$

The following theorems are useful for convergence analysis.

Theorem *(Doob, 1953). Let* $\{x_n, \mathcal{F}_n\}$ *be a nonnegative* $\left(x_n \overset{a.s.}{\geq} 0\right)$ *supermartingale such that*

$$\sup_n E\{x_n\} < \infty.$$

Then there exists a random variable x *(defined on the same probability space) such that*

$$E\{x\} < \infty, \qquad x_n \underset{n\to\infty}{\to} x \text{ (a.s.)}.$$

Theorem *(Robbins and Siegmund, 1971). Let* $\{\mathcal{F}_n\}$ *be a sequence of* $\sigma-$*algebras and* x_n, α_n, β_n *and* ξ_n *are* $\mathcal{F}_n$*-measurable nonnegative random variables such that for all* $n = 1, 2, ...$ *there exists* $E\{x_{n+1}/\mathcal{F}_n\}$ *and the following inequality is verified*

$$E\{x_{n+1}/\mathcal{F}_n\} \leq x_n(1+\alpha_n) + \beta_n - \xi_n$$

with probability one.
Then, for all $\omega \in \Omega_0$ *where*

$$\Omega_0 := \left\{\omega \in \Omega \middle| \sum_{n=1}^{\infty} \alpha_n < \infty, \ \sum_{n=1}^{\infty} \beta_n < \infty\right\}$$

the limit

$$\lim_{n\to\infty} x_n = x^*(\omega)$$

exists, and the sum

$$\sum_{n=1}^{\infty} \xi_n < \infty$$

converges.

Since the literature on stochastic processes is extensive we refer the reader to the books written respectively by Doob (1953) and Neveu (1975).

Index

Lecture Notes in Control and Information Sciences

Edited by M. Thoma

1993–1996 Published Titles:

Vol. 186: Sreenath, N.
Systems Representation of Global Climate Change Models. Foundation for a Systems Science Approach.
288 pp. 1993 [3-540-19824-5]

Vol. 187: Morecki, A.; Bianchi, G.; Jaworeck, K. (Eds)
RoManSy 9: Proceedings of the Ninth CISM-IFToMM Symposium on Theory and Practice of Robots and Manipulators.
476 pp. 1993 [3-540-19834-2]

Vol. 188: Naidu, D. Subbaram
Aeroassisted Orbital Transfer: Guidance and Control Strategies
192 pp. 1993 [3-540-19819-9]

Vol. 189: Ilchmann, A.
Non-Identifier-Based High-Gain Adaptive Control
220 pp. 1993 [3-540-19845-8]

Vol. 190: Chatila, R.; Hirzinger, G. (Eds)
Experimental Robotics II: The 2nd International Symposium, Toulouse, France, June 25-27 1991
580 pp. 1993 [3-540-19851-2]

Vol. 191: Blondel, V.
Simultaneous Stabilization of Linear Systems
212 pp. 1993 [3-540-19862-8]

Vol. 192: Smith, R.S.; Dahleh, M. (Eds)
The Modeling of Uncertainty in Control Systems
412 pp. 1993 [3-540-19870-9]

Vol. 193: Zinober, A.S.I. (Ed.)
Variable Structure and Lyapunov Control
428 pp. 1993 [3-540-19869-5]

Vol. 194: Cao, Xi-Ren
Realization Probabilities: The Dynamics of Queuing Systems
336 pp. 1993 [3-540-19872-5]

Vol. 195: Liu, D.; Michel, A.N.
Dynamical Systems with Saturation Nonlinearities: Analysis and Design
212 pp. 1994 [3-540-19888-1]

Vol. 196: Battilotti, S.
Noninteracting Control with Stability for Nonlinear Systems
196 pp. 1994 [3-540-19891-1]

Vol. 197: Henry, J.; Yvon, J.P. (Eds)
System Modelling and Optimization
975 pp approx. 1994 [3-540-19893-8]

Vol. 198: Winter, H.; Nüßer, H.-G. (Eds)
Advanced Technologies for Air Traffic Flow Management
225 pp approx. 1994 [3-540-19895-4]

Vol. 199: Cohen, G.; Quadrat, J.-P. (Eds)
11th International Conference on Analysis and Optimization of Systems – Discrete Event Systems: Sophia-Antipolis, June 15–16–17, 1994
648 pp. 1994 [3-540-19896-2]

Vol. 200: Yoshikawa, T.; Miyazaki, F. (Eds)
Experimental Robotics III: The 3rd International Symposium, Kyoto, Japan, October 28-30, 1993
624 pp. 1994 [3-540-19905-5]

Vol. 201: Kogan, J.
Robust Stability and Convexity
192 pp. 1994 [3-540-19919-5]

Vol. 202: Francis, B.A.; Tannenbaum, A.R. (Eds)
Feedback Control, Nonlinear Systems, and Complexity
288 pp. 1995 [3-540-19943-8]

Vol. 203: Popkov, Y.S.
Macrosystems Theory and its Applications: Equilibrium Models
344 pp. 1995 [3-540-19955-1]

Vol. 204: Takahashi, S.; Takahara, Y.
Logical Approach to Systems Theory
192 pp. 1995 [3-540-19956-X]

Vol. 205: Kotta, U.
Inversion Method in the Discrete-time Nonlinear Control Systems Synthesis Problems
168 pp. 1995 [3-540-19966-7]

Vol. 206: Aganovic, Z.;.Gajic, Z.
Linear Optimal Control of Bilinear Systems with Applications to Singular Perturbations and Weak Coupling
133 pp. 1995 [3-540-19976-4]

Vol. 207: Gabasov, R.; Kirillova, F.M.; Prischepova, S.V.
Optimal Feedback Control
224 pp. 1995 [3-540-19991-8]

Vol. 208: Khalil, H.K.; Chow, J.H.; Ioannou, P.A. (Eds)
Proceedings of Workshop on Advances inControl and its Applications
300 pp. 1995 [3-540-19993-4]

Vol. 209: Foias, C.; Özbay, H.; Tannenbaum, A.
Robust Control of Infinite Dimensional Systems: Frequency Domain Methods
230 pp. 1995 [3-540-19994-2]

Vol. 210: De Wilde, P.
Neural Network Models: An Analysis
164 pp. 1996 [3-540-19995-0]

Vol. 211: Gawronski, W.
Balanced Control of Flexible Structures
280 pp. 1996 [3-540-76017-2]

Vol. 212: Sanchez, A.
Formal Specification and Synthesis of Procedural Controllers for Process Systems
248 pp. 1996 [3-540-76021-0]

Vol. 213: Patra, A.; Rao, G.P.
General Hybrid Orthogonal Functions and their Applications in Systems and Control
144 pp. 1996 [3-540-76039-3]

Vol. 214: Yin, G.; Zhang, Q. (Eds)
Recent Advances in Control and Optimization of Manufacturing Systems
240 pp. 1996 [3-540-76055-5]

Vol. 215: Bonivento, C.; Marro, G.; Zanasi, R. (Eds)
Colloquium on Automatic Control
240 pp. 1996 [3-540-76060-1]

Vol. 216: Kulhavý, R.
Recursive Nonlinear Estimation: A Geometric Approach
244 pp. 1996 [3-540-76063-6]

Vol. 217: Garofalo, F.; Glielmo, L. (Eds)
Robust Control via Variable Structure and Lyapunov Techniques
336 pp. 1996 [3-540-76067-9]

Vol. 218: van der Schaft, A.
L_2 Gain and Passivity Techniques in Nonlinear Control
176 pp. 1996 [3-540-76074-1]

Vol. 219: Berger, M.-O.; Deriche, R.; Herlin, I.; Jaffré, J.; Morel, J.-M. (Eds)
ICAOS '96: 12th International Conference on Analysis and Optimization of Systems - Images, Wavelets and PDEs:
Paris, June 26-28 1996
378 pp. 1996 [3-540-76076-8]

Vol. 220: Brogliato, B.
Nonsmooth Impact Mechanics: Models, Dynamics and Control
420 pp. 1996 [3-540-76079-2]

Vol. 221: Kelkar, A.; Joshi, S.
Control of Nonlinear Multibody Flexible Space Structures
160 pp. 1996 [3-540-76093-8]

Vol. 222: Morse, A.S.
Control Using Logic-Based Switching
288 pp. 1997 [3-540-76097-0]

Vol. 223: Khatib, O.; Salisbury, J.K.
Experimental Robotics IV: The 4th International Symposium, Stanford, California, June 30 - July 2, 1995
596 pp. 1997 [3-540-76133-0]

Vol. 224: Magni, J.-F.; Bennani, S.;
Terlouw, J. (Eds)
Robust Flight Control: A Design Challenge
664 pp. 1997 [3-540-76151-9]